AQA Science
Further Additional Science

Revision Guide

GCSE

Pauline Anning
Niva Miles
John Scottow

Editor
Lawrie Ryan

OXFORD
UNIVERSITY PRESS

Great Clarendon Street, Oxford, OX2 6DP, United Kingdom

Oxford University Press is a department of the University of Oxford.
It furthers the University's objective of excellence in research, scholarship,
and education by publishing worldwide. Oxford is a registered trade mark of
Oxford University Press in the UK and in certain other countries

First published by Nelson Thornes Ltd in 2013
This edition published by Oxford University Press in 2014

British Library Cataloguing in Publication Data
Data available

978-1-4085-2428-2

10 9 8 7 6 5 4 3

Printed in China

Acknowledgements
Cover photograph: Fotolia
Page make-up: Wearset Ltd, Boldon, Tyne and Wear

Although we have made every effort to trace and contact all
copyright holders before publication this has not been possible in all
cases. If notified, the publisher will rectify any errors or omissions at
the earliest opportunity.

Links to third party websites are provided by Oxford in good faith
and for information only. Oxford disclaims any responsibility for
the materials contained in any third party website referenced in
this work.

Further Additional Science | Contents

Welcome to AQA GCSE Science!	**1**

B3 Unit 3 Biology — 2

1 Exchange of materials — **2**
1.1 Osmosis — 2
1.2 Active transport — 2
1.3 The sports drink dilemma — 3
1.4 Exchanging materials – the lungs — 3
1.5 Ventilating the lungs — 4
1.6 Artificial breathing aids — 5
1.7 Exchange in the gut — 6
1.8 Exchange in plants — 6
1.9 Transpiration — 7
End of chapter questions — 8

2 Transporting materials — **9**
2.1 The circulatory system and the heart — 9
2.2 Keeping the blood flowing — 10
2.3 Transport in the blood — 11
2.4 Artificial or real? — 12
2.5 Transport systems in plants — 12
End of chapter questions — 13

3 Keeping internal conditions constant — **14**
3.1 Controlling internal conditions — 14
3.2 The human kidney — 14
3.3 Dialysis – an artificial kidney — 15
3.4 Kidney transplants — 16
3.5 Controlling body temperature — 17
3.6 Treatment and temperature issues — 18
3.7 Controlling blood glucose — 18
3.8 Treating diabetes — 19
End of chapter questions — 20

4 How humans can affect the environment — **21**
4.1 The effects of the population explosion — 21
4.2 Land and water pollution — 22
4.3 Air pollution — 22
4.4 Deforestation and peat destruction — 23
4.5 Global warming — 24
4.6 Biofuels — 25
4.7 Biogas — 26
4.8 Making food production efficient — 26
4.9 Sustainable food production — 27
4.10 Environmental issues — 28
End of chapter questions — 29

Practice questions — **30**

C3 Unit 3 Chemistry — 32

1 The periodic table — **32**
1.1 The early periodic table — 32
1.2 The modern periodic table — 33
1.3 Group 1 – the alkali metals — 34
1.4 The transition elements — 35
1.5 Group 7 – the halogens — 36
End of chapter questions — 37

2 Water — **38**
2.1 Hard water — 38
2.2 Removing hardness — 39
2.3 Water treatment — 40
2.4 Water issues — 40
End of chapter questions — 41

3 Energy calculations — **42**
3.1 Comparing the energy released by fuels — 42
3.2 Energy transfers in solutions — 43
3.3 Energy level diagrams — 44
3.4 Calculations using bond energies — 45
3.5 Fuel issues — 46
End of chapter questions — 47

4 Analysis and synthesis — **48**
4.1 Tests for positive ions — 48
4.2 Tests for negative ions — 49
4.3 Titrations — 50
4.4 Titration calculations — 51
4.5 Chemical analysis — 52
4.6 Chemical equilibrium — 53
4.7 Altering conditions — 54
4.8 Making ammonia – the Haber process — 55
4.9 The economics of the Haber process — 55
End of chapter questions — 56

5 Organic chemistry — **57**
5.1 Structures of alcohols, carboxylic acids and esters — 57
5.2 Properties and uses of alcohols — 58
5.3 Carboxylic acids and esters — 59
5.4 Organic issues — 60
End of chapter questions — 61

Practice questions — **62**

P3 Unit 3 Physics   64

1 Medical applications of physics   64
1.1 X-rays   64
1.2 Ultrasound   64
1.3 Refractive index   65
1.4 The endoscope   66
1.5 Lenses   67
1.6 Using lenses   68
1.7 The eye   69
1.8 More about the eye   70
End of chapter questions   71

2 Using physics to make things work   72
2.1 Moments   72
2.2 Centre of mass   73
2.3 Moments in balance   74
2.4 Stability   74
2.5 Hydraulics   75
2.6 Circular motion   76
2.7 The pendulum   77
End of chapter questions   78

3 Using magnetic fields to keep things moving   79
3.1 Electromagnets   79
3.2 The motor effect   80
3.3 Electromagnetic induction   81
3.4 Transformers   82
3.5 Transformers in action   83
3.6 A physics case study   84
End of chapter questions   85

Practice questions   86

Answers   88
Glossary   94
Acknowledgements   98

Welcome to AQA GCSE Science!

Key points

At the start of each topic are the important points that you must remember.

This book has been written for you by experienced teachers and subject experts. Using this book will help you to prepare for your exams and is packed full of features to help you achieve the very best that you can.

Key words are highlighted in the text and are shown **like this**. You can look them up in the glossary at the back of the book if you're not sure what they mean.

Where you see this icon, you will know that this topic involves How Science Works – a really important part of your GCSE.

These questions check that you understand what you're learning as you go along. The answers are all at the back of the book.

Many diagrams are as important for you to learn as the text, so make sure you revise them carefully.

Anything in the Higher boxes must be learned by those sitting the Higher Tier exam. If you're sitting the Foundation Tier, these boxes can be missed out.

The same is true for any other places that are marked [**H**].

Higher

Study tip

Hints to help your study and exam preparation.

Bump up your grade

How you can improve your grade – this feature shows you where additional marks can be gained.

Maths skills

This feature highlights the maths skills that you will need for your Science exams with short, visual explanations.

At the end of each chapter you will find:

End of chapter questions

These questions will test you on what you have learned throughout the whole chapter, helping you to work out what you have understood and where you need to go back and revise.

And at the end of each unit you will find:

Practice questions

These are examples of the types of questions you may encounter in your GCSE exam, so you can get lots of practice during your course.

You can find answers to the End of chapter and Practice questions at the back of the book.

AQA Examination questions are reproduced by permission of AQA Education (AQA).

Student Book
pages 214–215 **B3**

1.1 Osmosis

- **Osmosis** is the movement of water.
- Just like diffusion, the movement of water molecules is random and requires no energy from the cell.
- Osmosis is the diffusion of water across a **partially permeable** membrane.
- The water moves from a region of high water concentration (a **dilute** solution) to a region of lower water concentration (a more **concentrated** solution).
- The cell membrane is partially permeable.

▶ **1** *What type of membrane is the cell membrane?*

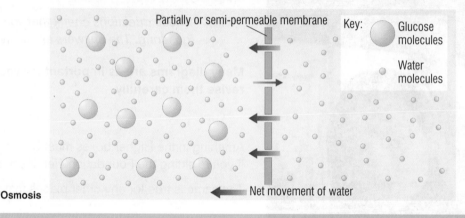

Osmosis

Key words: osmosis, partially permeable

Key points

- Osmosis is the diffusion of water.
- Water diffuses from a dilute solution to a more concentrated solution through a partially permeable membrane.
- Water moves into or out of cells by osmosis through the cell membrane.

Study tip

Remember that osmosis refers to the movement of water molecules only.

Student Book
pages 216–217 **B3**

1.2 Active transport

- Cells may need to absorb substances which are in short supply, i.e. against the concentration gradient.
- Cells use **active transport** to absorb substances across partially permeable membranes against the concentration gradient.
- Active transport requires the use of energy released in respiration.
- Cells are able to absorb ions from dilute solutions. For example, root cells absorb mineral ions from the dilute solutions in the soil by active transport.
- Glucose can be reabsorbed in the kidney tubules by active transport.

▶ **1** *Why does active transport require energy?*

Transport protein rotates and releases molecule inside cell (using energy)

Transport protein rotates back again (often using energy)

Useful molecule

Transport protein

| Outside cell | Inside cell | Outside cell | Inside cell | Outside cell | Inside cell |

Active transport uses energy to move substances against a concentration gradient

Key word: active transport

Key points

- Active transport allows cells to take in substances against a concentration gradient.
- Active transport uses energy from respiration.
- Ions, from dilute solutions, and other substances such as sugar can be absorbed by active transport.

Study tip

Remember that active transport needs energy; diffusion and osmosis do not.

1.3 The sports drink dilemma

- When you exercise, your muscles respire to release energy. Glucose, a sugar, is used in respiration.
- After a while, your body gets hot and you sweat more to cool down. Sweat contains water and mineral ions. If you sweat a lot your body cells may become **dehydrated**.
- If you exercise for a long time it may be necessary to replace the sugar, mineral ions and water which have been used or lost.

> **1** *Why is glucose used during exercise?*

- Sports drinks are solutions of sugar and mineral ions.
- The water in the drinks helps to **rehydrate** body cells.
- The drinks are designed to help balance the concentration of body fluids and the concentrations inside cells. If the drink concentration matches the body fluids the solution is called **isotonic**.
- The evidence for the benefits of sports drinks is varied. Some sports scientists think it is just as good to have a drink of water for short periods of exercise. The drinks may help athletes who need to replace mineral ions and sugar as well as water.

> **2** *How are mineral ions lost from the body during exercise?*

Key words: dehydrated, rehydrate, isotonic

Key points

- Most soft drinks contain:
  - water to replace water lost when sweating,
  - sugar to replace sugar used for energy release in exercise,
  - mineral ions to replace those lost in sweating.
- For normal levels of exercise, water is likely to be as effective as a sports drink.

Bump up your grade

Questions on sports drinks often have data in the form of graphs and tables. To make sure you get all the available marks, quote figures from the data when you are asked to explain why something is relevant.

1.4 Exchanging materials – the lungs

- Large, complex, organisms have special **exchange surfaces** to obtain all the food and oxygen they need. Soluble food materials (**solutes**) are absorbed by the intestine. Oxygen is absorbed by the lungs and carbon dioxide is removed from them.
- Efficient exchange surfaces have a large surface area, thin walls or a short diffusion path, and an efficient transport system – the blood supply in animals.
- The lungs contain the **gaseous exchange** surface. The surface area of the lungs is increased by the **alveoli** (air sacs).
- The alveoli have thin walls, a large surface area and a good blood supply.
- The lungs are **ventilated** to maintain a steep diffusion gradient.
- Oxygen diffuses into the many **capillaries** surrounding the alveoli and carbon dioxide diffuses back out into the lungs to be breathed out.

> **1** *Which structures increase the surface area of the lungs?*

Study tip

Make sure you can explain how the features of an exchange surface make them efficient.

Key words: exchange surface, solute, gaseous exchange, alveoli, ventilated, capillary

Key points

- Large organisms need exchange surfaces.
- Exchange surfaces have features to make them efficient.
- The lungs are adapted to be an efficient exchange surface.

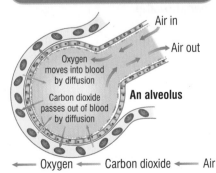

The alveoli are adapted so that gas exchange can take place as efficiently as possible in the lungs

Student Book
pages 222–223
B3

1.5 Ventilating the lungs

- The lungs contain the exchange surface of the **breathing system**.
- The lungs are situated in the thorax, inside the ribcage and above the **diaphragm**, which separates the lungs from the **abdomen**.
- When we **breathe** in:
  - The **intercostal muscles**, between the ribs and the diaphragm, contract.
  - The ribcage moves up and out and the diaphragm flattens.
  - The volume of the thorax increases.
  - The pressure in the thorax decreases and air is drawn in.
- When we breathe out:
  - The intercostal muscles of the ribcage and diaphragm relax.
  - The ribcage moves down and in and the diaphragm becomes **domed**.
  - The volume of the thorax decreases.
  - The pressure increases and air is forced out.
- The movement of air in and out of the lungs is known as ventilation.

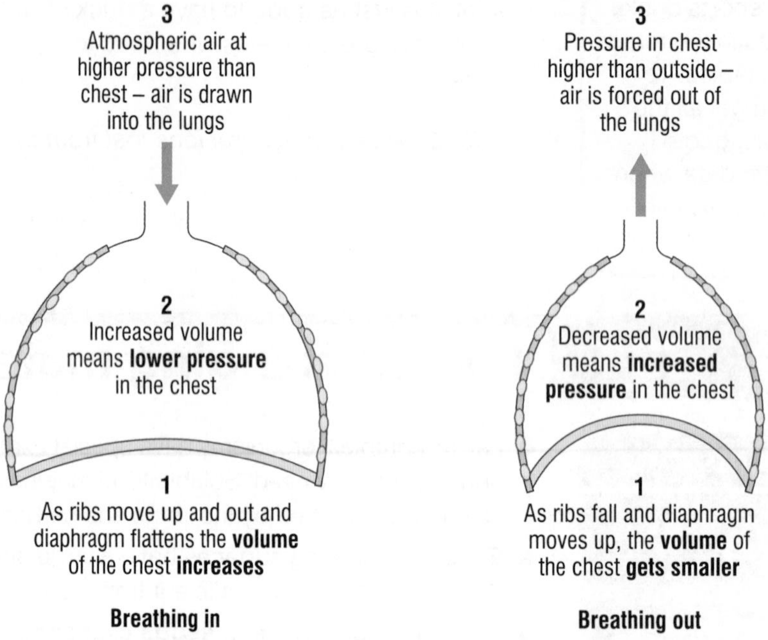

Key points

- The breathing system is designed to move air in and out of the lungs.
- Ventilation occurs due to changes of pressure inside the chest.

Study tip

Make sure you understand how breathing movements alter the volume and pressure inside the thorax.

3
Atmospheric air at higher pressure than chest – air is drawn into the lungs

2
Increased volume means **lower pressure** in the chest

1
As ribs move up and out and diaphragm flattens the **volume** of the chest **increases**

Breathing in

3
Pressure in chest higher than outside – air is forced out of the lungs

2
Decreased volume means **increased pressure** in the chest

1
As ribs fall and diaphragm moves up, the **volume** of the chest **gets smaller**

Breathing out

Ventilation of the lungs

1 *Which muscles contract when you breathe in?*

Key words: breathing system, diaphragm, abdomen, breathe, intercostal muscles, domed

Student Book
pages 224–225

B3

1.6 Artificial breathing aids

Key points

- There are advantages and disadvantages when using different types of artificial breathing aids.

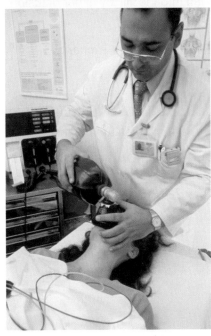

Using a positive pressure bag ventilator

There are many reasons why someone cannot get enough oxygen into their bloodstream:

- If the alveoli are damaged, the surface area for gas exchange is reduced.
- If the tubes leading to the lungs are narrowed, less air can be moved through them.
- If the person is paralysed, their muscles will not work to pull the ribcage up and out.

Several types of breathing aid have been developed:

- The 'iron lung' was used for people with polio who were paralysed. The person lay with their chest sealed in a large metal cylinder. When air was drawn out of the cylinder the person's chest moved out and they breathed in. The **vacuum** which was formed inside the cylinder created a **negative pressure**. When air was pumped back in to the cylinder it created pressure on the chest and forced air out of the person's lungs.
- Breathing aids which force measured amounts of air into the lungs use **positive pressure**. Bags of air linked to masks can force air down the **trachea**.
- Positive pressure aids are often smaller, easier to manage in the home and can be linked to computers for control.

⟱➡ **1** *Give one advantage of a positive pressure breathing aid compared with the iron lung.*

Study tip

Try to find examples of artificial breathing aids on the internet and write down their advantages or disadvantages.

Key words: vacuum, negative pressure, positive pressure, trachea

Student Book
pages 226–227 **B3**

1.7 Exchange in the gut

Key points

- The villi are an efficient exchange surface for the absorption of the soluble products of digestion by diffusion or active transport.

- Villi provide a large surface area and have a good blood supply.

- The food we eat is digested in the gut into small, soluble molecules. In the small intestine these solutes are absorbed into the blood. The **villi** line the inner surface of the small intestine and are the exchange surface for food molecules.

- The villi are finger-like projections which greatly increase the surface area for absorption to take place.

- The walls of the villi are very thin and there are many capillaries close to the wall.

- The soluble products of digestion can be absorbed into the villi by either diffusion or active transport.

▷ **1** *Give two features of the villi which make them efficient exchange surfaces.*

◁ Bump up your grade

To make sure you get maximum marks, always relate the features of an exchange surface to its function. The two key examples to learn are alveoli (in the lungs) and villi (in the small intestine).

Key word: villus

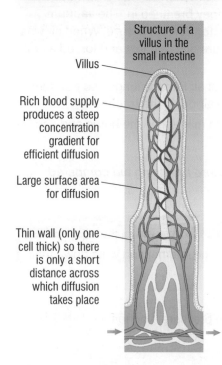

Structure of a villus in the small intestine

Villus

Rich blood supply produces a steep concentration gradient for efficient diffusion

Large surface area for diffusion

Thin wall (only one cell thick) so there is only a short distance across which diffusion takes place

Thousands of finger-like projections in the wall of the small intestine – the villi – make it possible for all the digested food molecules to be transferred from your small intestine into your blood by diffusion and active transport

Student Book
pages 228–229 **B3**

1.8 Exchange in plants

Key points

- The stomata in plant leaves allow carbon dioxide to diffuse into the leaf from the atmosphere.

- Leaves are flat and thin with internal air spaces to increase the surface area for diffusion of gases.

- Root hair cells increase the surface area of roots.

- Gases diffuse in and out of leaves through tiny holes called 'stomata'. The size of the stomata is controlled by guard cells which surround them. These gases are:

  - Oxygen: needed for respiration and is a waste product of photosynthesis.

  - Carbon dioxide: needed for photosynthesis and is a waste product of respiration. The movement of these gases depends upon which process is taking place the most quickly.

- Plants also lose water vapour through the stomata due to **evaporation** in the leaves.

- Leaves are flat and very thin so the gases do not need to diffuse very far. There are also internal air spaces.

- Water and mineral ions are taken up by the roots. **Root hair cells** increase the surface area of roots for the absorption of water and mineral ions.

- If plants lose water faster than it is replaced by the roots, the stomata can close to prevent wilting.

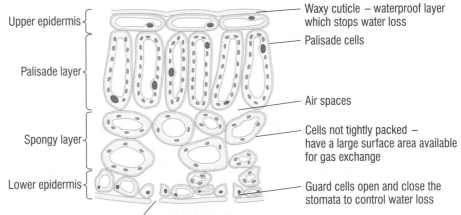

Upper epidermis — Waxy cuticle – waterproof layer which stops water loss

Palisade cells

Palisade layer

Air spaces

Spongy layer

Cells not tightly packed – have a large surface area available for gas exchange

Lower epidermis

Guard cells open and close the stomata to control water loss

Stomata like this allow gases to move in and out of the leaf

This is a cross section of a leaf showing the arrangement of the cells inside, with plenty of air spaces and short diffusion distances. This means that the carbon dioxide needed for photosynthesis reaches the cells as efficiently as possible.

> **1** *Why do leaves have stomata?*

Key words: evaporation, root hair cell

Student Book
pages 230–231 **B3**

1.9 Transpiration

- Plants take up water through the roots. The water passes through the plant to the leaves. In the leaves the water evaporates from the leaf cells and the water vapour diffuses out through the stomata.
- The movement of the water through the plant is called the **transpiration stream**.
- The plant could dehydrate if the rate of evaporation in the leaves is greater than the water uptake by the roots.
- Evaporation is more rapid in hot, dry, windy or bright conditions.
- The **guard cells** can close to prevent excessive water loss. **Wilting** of the whole plant can also reduce water loss. The leaves collapse and hang down, which reduces the surface area.

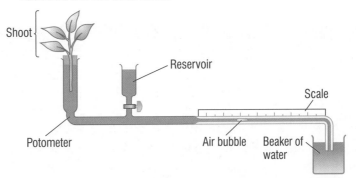

Shoot

Reservoir

Scale

Potometer

Air bubble

Beaker of water

A potometer can be used to show how the uptake of water by a plant changes with different conditions. This gives you a good idea of the amount of water lost by transpiration.

> **1** *By what process does water vapour move out of the leaves?*

Key words: transpiration stream, guard cells, wilting

Key points

- Water is lost from the leaves by evaporation.
- The water vapour escapes the leaf through the stomata when they open to allow carbon dioxide to enter the leaf.
- Wilting can protect the leaf from excessive water loss.

Study tip

Conditions which increase the rate of evaporation will increase the rate of transpiration.

1 In what way is osmosis different from diffusion?

2 Give an example of when active transport is carried out by cells.

3 What is meant by 'against a concentration gradient'?

4 Name the exchange surfaces in the lungs and intestine.

5 What are the features of an efficient exchange surface?

6 Describe the process involved in breathing in.

7 Which sort of patients benefit from an iron lung machine?

8 How is the surface area of a root increased and why is this important?

9 Why is the rate of transpiration higher on a bright day?

10 On humid days, the rate of transpiration is slow even when it is hot. Explain why.

Chapter checklist ✔ ✔ ✔

Tick when you have:

reviewed it after your lesson	✔ ☐ ☐
revised once – some questions right	✔ ✔ ☐
revised twice – all questions right	✔ ✔ ✔

Move on to another topic when you have all three ticks

Osmosis	☐ ☐ ☐
Active transport	☐ ☐ ☐
The sports drink dilemma	☐ ☐ ☐
Exchanging materials – the lungs	☐ ☐ ☐
Ventilating the lungs	☐ ☐ ☐
Artificial breathing aids	☐ ☐ ☐
Exchange in the gut	☐ ☐ ☐
Exchange in plants	☐ ☐ ☐
Transpiration	☐ ☐ ☐

2.1 The circulatory system and the heart

Key points

- The double circulation system in humans consists of blood vessels, blood and the heart.
- The heart is the organ that pumps blood around the body.
- Valves prevent backflow of blood.

Study tip

Practise labelling diagrams of the heart. You are often asked to identify the chambers and the main blood vessels. Remember the sequence: veins → atria → ventricles → arteries

Large organisms need a **transport system** to move materials around the body.

- Humans have a **circulatory system** which consists of blood vessels, the **heart** and blood.
- The heart is a muscular organ that pumps blood around the body. It is actually two pumps held together.
- The right pump forces **deoxygenated** blood to the lungs where it picks up oxygen and loses carbon dioxide.
- After returning to the heart, the **oxygenated** blood is then pumped to the rest of the body by the left pump.
- The heart has four chambers. The upper ones called **atria** receive blood from the **vena cava** on the right and **pulmonary vein** on the left. The atria contract to move blood into the lower chambers, the **ventricles**. When the ventricles contract they force blood into the **pulmonary artery** from the right side and into the **aorta** on the left side. Valves in the heart prevent the blood from flowing in the wrong direction. The heart muscle is supplied with oxygenated blood via the **coronary arteries**.
- The action of the two sides of the heart results in a **double circulation**.

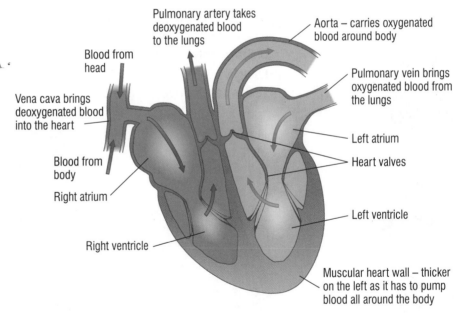

The structure of the heart

Labels in diagram:
- Pulmonary artery takes deoxygenated blood to the lungs
- Aorta – carries oxygenated blood around body
- Blood from head
- Vena cava brings deoxygenated blood into the heart
- Pulmonary vein brings oxygenated blood from the lungs
- Left atrium
- Heart valves
- Blood from body
- Right atrium
- Left ventricle
- Right ventricle
- Muscular heart wall – thicker on the left as it has to pump blood all around the body

Handwritten notes:
Vena Cava → carbonated.
vein
→ mgre nor red wbod
Aorta
I ought to sub blood to body

▐▐▶ **1** *What is the function of valves in the heart?*

Key words: transport system, heart, deoxygenated, oxygenated, atria, vena cava, pulmonary vein, ventricles, pulmonary artery, aorta, coronary artery, double circulation

2.2 Keeping the blood flowing

Blood flows round the body in three main types of blood vessel: the arteries, veins and capillaries:

Arteries
- carry blood away from the heart
- have thick walls containing muscle and elastic tissue.

Veins
- have thinner walls than arteries
- often have **valves** along their length to prevent backflow of blood.

Capillaries
- are narrow, thin-walled vessels
- carry the blood through the organs and allow the exchange of substances with all the living cells in the body.

> **1** *What is the difference in structure between an artery and vein?*

The heart keeps the blood flowing through the blood vessels. If blood vessels are blocked or too narrow the blood will not flow efficiently. Then organs will be deprived of nutrients and oxygen.

- **Stents** can be inserted to keep blood vessels open. This is particularly beneficial when coronary arteries become narrowed due to fatty deposits, cutting off the blood supply to the heart muscle.
- Leaky valves mean the blood could flow in the wrong direction. Artificial or animal valves can be inserted in the heart to replace damaged valves.

Key points

- Arteries, veins and capillaries are the main types of blood vessel.
- Substances diffuse between the blood and the cells in the capillaries.
- Blocked or narrow arteries can be widened by stents.
- Damaged heart valves can be replaced.

Study tip

Make sure you can explain why a narrowed blood vessel or leaky valve causes problems for a person's health.

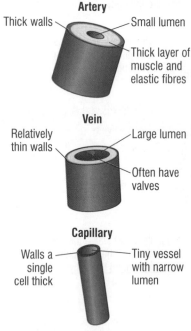

Artery
Thick walls — Small lumen
— Thick layer of muscle and elastic fibres

Vein
Relatively thin walls — Large lumen
— Often have valves

Capillary
Walls a single cell thick — Tiny vessel with narrow lumen

The three main types of blood vessel

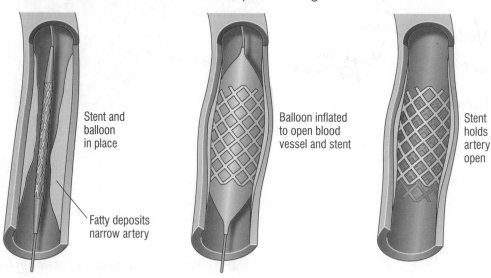

Stent and balloon in place

Balloon inflated to open blood vessel and stent

Stent holds artery open

Fatty deposits narrow artery

A stent being positioned in an artery

> **2** *Why are stents used?*

Key words: valve, stent

2.3 Transport in the blood

Blood is a tissue. The fluid **plasma** contains **red blood cells**, **white blood cells** and **platelets**.

Key points

- Blood plasma transports dissolved food molecules, carbon dioxide and urea.
- Plasma also contains blood cells.
- The red blood cells carry oxygen from the lungs to the cells.
- The white cells are part of the defence system.
- Platelets are involved in blood clotting.

- Blood plasma transports many substances including:
  - carbon dioxide from the organs to the lungs
  - soluble products of digestion from the small intestine to other organs
  - **urea** from the liver to the kidneys where **urine** is made.

▶ **1 Give an example of a gas transported by plasma.**

- Red blood cells:
  - are **biconcave discs** which do not have a nucleus
  - contain the red **pigment haemoglobin**
  - use their haemoglobin which combines with oxygen to form **oxyhaemoglobin** in the lungs
  - carry the oxygen to all the organs where the oxyhaemoglobin splits into haemoglobin and oxygen.
- White blood cells:
  - have a nucleus
  - form part of the body's defence system against microorganisms.
- Platelets:
  - are small fragments of cells
  - do not have a nucleus
  - help blood to clot at the site of a wound.

▶ **2 What is the function of platelets?**

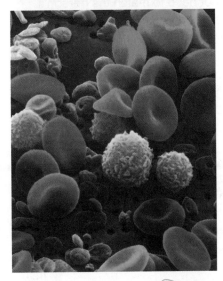

Red blood cell   Platelets   White blood cell

Red blood cells, white blood cells and platelets are suspended in the plasma to make up our blood

Study tip

Remember the function of a cell describes what it does – its job.
Do not confuse *function* with *structure*.

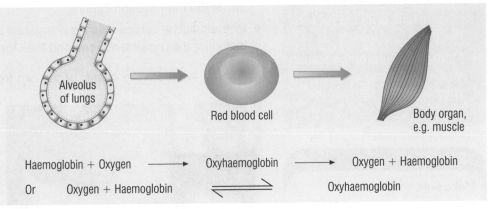

Alveolus of lungs → Red blood cell → Body organ, e.g. muscle

Haemoglobin + Oxygen ⟶ Oxyhaemoglobin ⟶ Oxygen + Haemoglobin

Or   Oxygen + Haemoglobin ⇌ Oxyhaemoglobin

The reversible reaction between oxygen and haemoglobin makes life as we know it possible by carrying oxygen to all the places where it is needed

Key words: plasma, red blood cell, platelet, urea, biconcave disc, pigment, haemoglobin, oxyhaemoglobin

B3

2.4 Artificial or real?

- Blood from blood **donors** can be separated into cells and plasma. Donated blood must be refrigerated. Some blood products can be frozen.
- The plasma can be given to patients in a **transfusion** to increase blood volume.
- Artificial blood such as **perfluorocarbons**, PFCs, may be used which does not need to be refrigerated. It does not contain cells so blood matching is not necessary.
- Artificial blood is expensive and does not carry as much oxygen as whole blood. Some types are insoluble in water so do not mix well with blood.
- Some artificial bloods cause unpleasant side-effects.

> **1** *Why is plasma given to patients?*

- Patients who suffer heart failure may need a new heart.
- There is a lack of heart donors so artificial hearts are being developed to keep the patients alive.
- Advantages of artificial hearts are that they do not need to match the person's tissue and there is no need for immunosuppressant drugs.
- Disadvantages of artificial hearts are problems with blood clotting, long stays in hospital and expense.

> **2** *What are the advantages of an artificial heart?*

Key words: donor, transfusion, perfluorocarbon

Key points

- There are advantages and disadvantages when using artificial blood or artificial hearts.
- They are useful when real blood or donor hearts are not available.

Bump up your grade

To gain maximum marks, don't be put off if there is an unfamiliar diagram in an exam question. Artificial aids must mimic the real thing. Think about the functions of a real heart or real blood.

B3

2.5 Transport systems in plants

- Flowering plants have separate transport systems.
- **Xylem** tissue transports water and mineral ions from the roots to the stem, leaves and flowers.
- The movement of water from the roots through the xylem and out of the leaves is called the transpiration stream.
- **Phloem** tissue carries dissolved sugars from the leaves to the rest of the plant, including the growing regions and the storage organs.

> **1** *What are the names of the two transport tissues in flowering plants?*

Key points

- Flowering plants have two transport tissues.
- Xylem transports water and mineral ions from the roots to the stems and leaves.
- Phloem transports food, such as dissolved sugar, from the leaves to other parts of the plant.

Study tip

Make sure you learn:
- xylem for water and mineral ion transport
- phloem for food transport.

A simple way of demonstrating that water moves up the xylem in celery

Key words: xylem, phloem

1 Which type of blood vessel exchanges materials with the cells?

2 What is the name of the red pigment in red blood cells?

3 Which blood vessels carry blood away from the heart?

4 Why does blood only flow in one direction?

5 What is the function of white blood cells?

6 What is transported by a) xylem, b) phloem?

7 Why do humans have a double circulation?

8 Some patients are fitted with a stent in a coronary artery. Explain why.

9 Why has it been necessary to develop artificial hearts?

10 If you were designing an artificial heart, what key features would you include?

Chapter checklist

Tick when you have:				The circulatory system and the heart			
reviewed it after your lesson	✓	☐	☐	Keeping the blood flowing	☐	☐	☐
revised once – some questions right	✓	✓	☐	Transport in the blood	☐	☐	☐
revised twice – all questions right	✓	✓	✓	Artificial or real?	☐	☐	☐
Move on to another topic when you have all three ticks				Transport systems in plants	☐	☐	☐

Student Book
pages 246–247 **B3**

3.1 Controlling internal conditions

Key points

- Waste products must be removed from the cells.
- Internal conditions in the body must be kept constant.

Study tip

Make sure you can explain what is meant by homeostasis.

The internal conditions of the body must be carefully controlled. Keeping the conditions within a narrow range is called **homeostasis**. Temperature, blood glucose, water and ion content and levels of waste products must all be controlled.

- Waste products that have to be removed from the body include:
  - carbon dioxide, produced by respiration, removed via the lungs when we breathe out
  - urea, produced in the **liver** from the breakdown of amino acids, removed by the kidneys in the urine and temporarily stored in the bladder.
- Water and ions enter the body when we eat and drink. If the water or ion content in the body is wrong, too much water may move into or out of the cells. This could damage or destroy the cells.

▶ **1** *Where is urea made?*

Key words: homeostasis, liver

Student Book
pages 248–249 **B3**

3.2 The human kidney

Key points

- Chemical reactions in the body produce substances that are toxic (poisonous) e.g. urea.
- The kidneys excrete substances that the body does not want.
- The kidneys first filter substances out of the blood. They then reabsorb the substances that the body needs.

Study tip

Remember that urea is made in the liver by the breakdown of amino acids. Urea is removed from the blood by the kidneys during filtration.

The kidney is a very important organ of homeostasis. It controls the balance of water and mineral ions in the body and gets rid of urea.

- The body has two kidneys. They **filter** the blood, excreting substances you do not want and keeping those substances that the body needs.
- A healthy kidney produces urine by:
  - first filtering the blood,
  - reabsorbing all the sugar,
  - reabsorbing the dissolved ions needed by the body,
  - reabsorbing as much water as the body needs,
  - releasing urea, excess ions and water in the urine.
- The urine is temporarily stored in the **bladder** before being removed from the body.

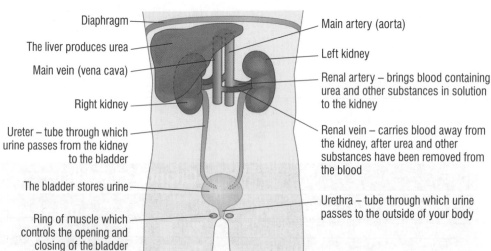

Diaphragm
The liver produces urea
Main vein (vena cava)
Right kidney
Ureter – tube through which urine passes from the kidney to the bladder
The bladder stores urine
Ring of muscle which controls the opening and closing of the bladder
Main artery (aorta)
Left kidney
Renal artery – brings blood containing urea and other substances in solution to the kidney
Renal vein – carries blood away from the kidney, after urea and other substances have been removed from the blood
Urethra – tube through which urine passes to the outside of your body

▶ **1** *What two processes occur in the kidney to make urine?*

Key word: bladder

Student Book
pages 250–251

B3

3.3 Dialysis – an artificial kidney

- If a person suffers from kidney failure they can be kept alive by **dialysis**.
- A **dialysis machine** carries out the same job as the kidneys. The blood flows between partially permeable membranes.
- The dialysis fluid contains the same concentration of useful substances that the patient's blood does, e.g. glucose and mineral ions. This means that these substances do not diffuse out of the blood so they do not need to be reabsorbed. Urea diffuses out from the blood into the dialysis fluid.
- Dialysis restores the concentration of substances in the blood back to normal, but needs to be carried out at regular intervals.
- If a kidney becomes available, the patient may have a **kidney transplant**. If the transplant is successful, the person will not need further dialysis.

Key points

- When kidneys fail, the patient can be treated by dialysis.
- A dialysis machine does the work of the kidneys and keeps patients alive.
- If a successful kidney transplant is carried out, then the dialysis machine will no longer be necessary.

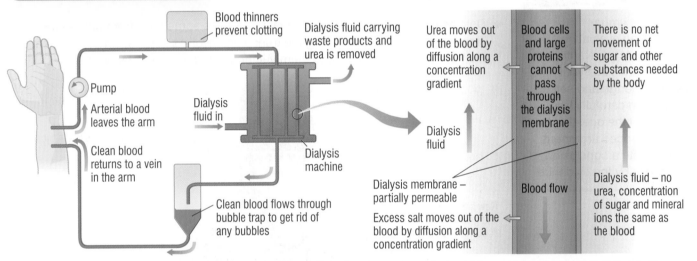

A dialysis machine relies on simple diffusion to clean the blood, removing the waste products which would damage the body if they built up

Study tip

Make sure you understand why the dialysis fluid contains glucose and mineral ions. If the dialysis fluid was water only, all the useful solutes would diffuse out of the blood as well as the urea.

▶ **1** *What is the purpose of dialysis?*

Key words: dialysis, dialysis machine, kidney transplant

3.4 Kidney transplants

- For most patients a kidney transplant is a better option than dialysis. The diseased kidney is replaced with a healthy one.
- Kidneys may be obtained from a victim of a fatal accident or sometimes from living donors.
- The new kidney must be a very good 'tissue match' to prevent rejection.
- There are proteins called antigens on the surface of cells. The **recipient**'s antibodies may attack the antigens on the donor organ because they recognise them as being 'foreign'.
- Following the transplant the recipient must take drugs to suppress the **immune response** to prevent rejection. These are called **immunosuppressant drugs**.

▌▌▶ **1** *What are antigens?*

- Despite the advantages of a transplant, there are some risks from operations. Treatment before and following the transplant involves suppressing the patient's immune system, which leaves them vulnerable to common infections.

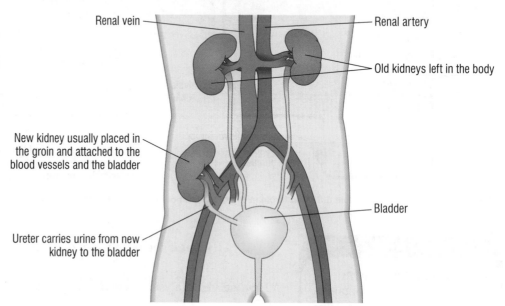

Renal vein

Renal artery

Old kidneys left in the body

New kidney usually placed in the groin and attached to the blood vessels and the bladder

Bladder

Ureter carries urine from new kidney to the bladder

A donor kidney takes over the functions of failed kidneys which are usually left in place

▌▌▶ **2** *What is the main advantage of a kidney transplant compared with dialysis?*

Key words: recipient, immune response, immunosuppressant drug

Key points

- A diseased kidney can be replaced by a healthy kidney.
- The donor kidney must be a very good 'match'.
- The immune system has to be 'suppressed' or it is likely to reject the new kidney.

Bump up your grade

To gain maximum marks when evaluating advantages/disadvantages of transplants, use your own knowledge and select information carefully from the stem of the question to give a balanced argument. Always write a conclusion backed up by the evidence.

Student Book
pages 254–255 **B3**

3.5 Controlling body temperature

Key points

- Enzymes work in a very narrow temperature range.
- Body temperature is monitored and controlled to keep it within this narrow range.

Study tip

Remember that the thermoregulatory centre in the brain detects **blood** temperature.

Human body temperature must be kept at about 37 °C so that the enzymes will work efficiently. The **core body temperature**, deep inside the body, must be kept stable.

- Body temperature is monitored and controlled by the **thermoregulatory centre** in the brain. This centre has receptors which detect the temperature of the blood flowing through the brain.
- Temperature receptors in the skin also send impulses to the brain to give information about skin temperature.
- The skin looks red when we are hot due to increased blood flow.
- Sweating helps to cool the body. When it is hot, more water is lost from the skin so more water must be taken in with drinks and food to balance this loss.

IIII▶ **1** *Where is blood temperature monitored?*

If the core temperature *rises*:

- Blood vessels near the surface of the skin dilate allowing more blood to flow through the skin capillaries. Energy is transferred by radiation and the skin cools.
- Sweat glands produce more sweat. Its water evaporates from the skin's surface. The energy required for the water to evaporate comes from the skin's surface. So we cool down.

If the core temperature *falls*:

- Blood vessels near the surface of the skin constrict and less blood flows through the skin capillaries. Less energy is radiated.
- We 'shiver'. Muscles contract quickly. This requires respiration and some of the energy released warms the blood.

Higher

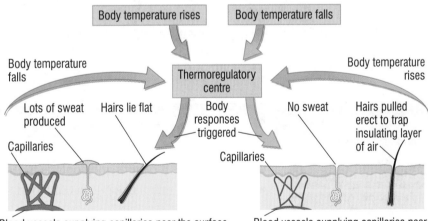

Maintaining a steady internal environment

IIII▶ **2** *Explain why sweating cools the body.* [H]

Key words: core body temperature, thermoregulatory centre

17

B3

3.6 Treatment and temperature issues ⚙️●

When doctors try to decide on the best course of treatment for patients with kidney failure they have to consider many issues. The two main treatments are dialysis and kidney transplants.

- The doctors need to consider:
  - the general health of the patient,
  - how long the patient has been on dialysis,
  - the total cost of treatment – the long term costs of continuing dialysis against an expensive operation followed by treatment with immunosuppressant drugs,
  - the risks of a transplant operation, e.g. risks associated with anaesthetics, infection,
  - the availability of donor kidneys.
- There are ethical issues too:
  - Should everyone be automatically on a transplant register or should they opt in?
  - Should people be paid to be donors?
  - Should people pay for a new kidney to jump the queue?
- Carrying a donor card means that a family does not have to make a distressing decision about organ donation immediately after the death of a close relative.

▶ **1** *Why is it important to carry a donor card or tell your family you want to be a donor?*

Extreme temperatures can be very dangerous for survival because the body's enzymes do not work properly.

Small children have a large surface area to volume ratio. This means they transfer energy to the surroundings very quickly in cold conditions and dehydrate very quickly in hot conditions. If the body temperature is too low the respiratory enzymes work too slowly and too little energy is released. If the child dehydrates they cannot cool down and they overheat, which means enzymes denature.

Elderly people suffer from hypothermia in cold conditions because they do not move around much to release energy from respiration in the muscles.

Explorers in extreme environments have to learn how to recognise the symptoms of hypothermia and dehydration.

B3

3.7 Controlling blood glucose

- The pancreas monitors and controls the level of glucose in our blood.
- If there is too much glucose in our blood the pancreas produces the hormone **insulin**.
- Insulin causes the glucose to move from the blood into the cells.
- In the liver, excess glucose is converted to **glycogen** for storage.
- If no, or too little, insulin is produced by the pancreas the blood glucose level may become very high. This condition is known as **Type 1 diabetes**.
- Type 1 diabetes is controlled by injections of insulin and careful attention to diet and levels of exercise.

▶ **1** *Which hormone reduces the level of glucose in the blood?*

Key words: insulin, Type 1 diabetes

- Insulin causes the blood glucose level to fall.
- If the level gets too low receptors in the pancreas detect the low level.
- The pancreas releases glucagon, another hormone.
- The glucagon causes the glycogen in the liver to change into glucose.
- This glucose is released back into the blood.

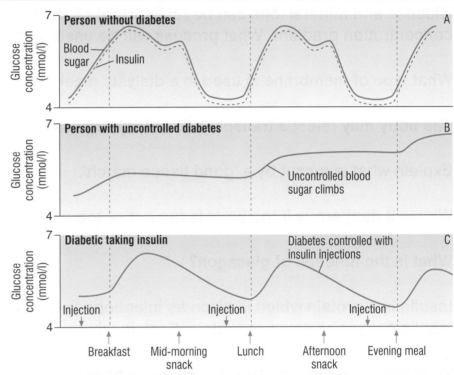

These graphs show the impact insulin injections have on people with Type 1 diabetes. The injections keep the blood glucose level within safe limits.

3.8 Treating diabetes

- Type 1 diabetes is traditionally treated with human insulin produced by genetically engineered bacteria.
- The diabetic has to inject before meals every day of their life.
- Very active diabetics have to match the amount of insulin injected with their diet and exercise.
- Some diabetics use pumps attached to the body. They can adjust the level of insulin injected by the pump.
- Doctors and other scientists are trying to develop new methods of treating and possibly curing Type 1 diabetes.

These include:
- pancreas transplants,
- transplanting pancreas cells,
- using **embryonic stem cells** to produce insulin secreting cells,
- using adult stem cells from diabetic patients,
- genetically engineering pancreas cells to make them work properly.

⟫ **1** *Name two new methods that are being developed to treat Type 1 diabetes.*

1 Why must urea be removed from the blood?

2 Suggest why the kidneys reabsorb glucose.

3 Glucose and mineral ions can be reabsorbed by the kidney against a concentration gradient. What process will be used to do this?

4 What type of membrane is used in a dialysis machine?

5 The body may reject a transplanted kidney. How is rejection prevented?

6 Explain what is meant by a 'good tissue match'.

7 Why is it dangerous if the body is too hot or too cold?

8 What is the function of glucagon? [H]

9 Insulin is a protein which is given as injections to people with Type 1 diabetes. Insulin cannot be given as tablets. Explain why.

10 What effect do diet and levels of exercise have on blood glucose levels?

Chapter checklist ✓ ✓ ✓

Tick when you have:

reviewed it after your lesson	✓ ☐ ☐	Controlling internal conditions	☐ ☐ ☐
revised once – some questions right	✓ ✓ ☐	The human kidney	☐ ☐ ☐
revised twice – all questions right	✓ ✓ ✓	Dialysis – an artificial kidney	☐ ☐ ☐
Move on to another topic when you have all three ticks		Kidney transplants	☐ ☐ ☐
		Controlling body temperature	☐ ☐ ☐
		Treatment and temperature issues	☐ ☐ ☐
		Controlling blood glucose	☐ ☐ ☐
		Treating diabetes	☐ ☐ ☐

Student Book
pages 264–265 **B3**

4.1 The effects of the population explosion

Key points

- The human population is increasing rapidly.
- Humans are using more resources and producing more waste and pollution.

Study tip

Don't forget that some problems arise miles away from the source of the pollution. For example, sulfur dioxide gas can be blown to another area before dissolving in rain.

- There are increasing numbers of people on our planet. Currently the world **population** is about 7 billion!
- Many people want and demand a better standard of living.
- We are using up raw materials and those that are **non-renewable** cannot be replaced.
- When goods are produced there is a lot of **industrial waste**.
- We are producing more waste and the pollution that goes with it.
- Humans reduce the amount of land available for animals and plants by building, quarrying, farming and dumping waste.
- The continuing increase in the human population is affecting the **ecology** of the Earth.

Humans pollute:
- waterways with **sewage**, **fertiliser** and toxic chemicals,
- air with smoke and gases such as sulfur dioxide, which contributes to acid rain,
- land with toxic chemicals such as **pesticides** and **herbicides** and these can then be washed into the water.

▶ **1** *Why does building houses for humans affect animals and plants?*

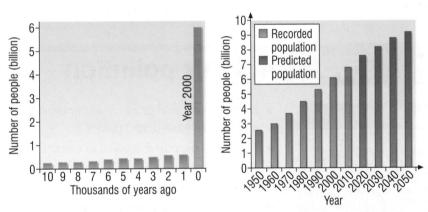

These records of human population growth shows the massive increase during the past years. The second bar chart predicts that this increase will continue.

Key words: non-renewable, industrial waste, ecology, sewage, fertiliser, pesticide, herbicide

Student Book pages 266–267 **B3**

4.2 Land and water pollution

Key points

- Humans are responsible for polluting both land and waterways.

Study tip

Make sure you know how humans pollute the land and water and the effect this has on other organisms.

People pollute land in many different ways:

- Sewage contains human body waste and waste water from our homes. Sewage must be treated properly to remove gut **parasites** and toxic chemicals or these can get onto the land.
- Large quantities of household and industrial waste are placed in landfill and toxic chemicals leak out. Some industrial waste, such as radioactive waste, is very hazardous.
- Farming methods can pollute the land.
- Herbicides (weedkillers) and pesticides (which kill insects) are also poisons. The poisons sprayed onto crops can get into the soil and into the food chain. Eventually many of them are washed into rivers.
- Farmers also use chemical fertilisers, to keep the soil **fertile**, which can be washed into rivers.

1 *Why does landfill cause pollution?*

Water pollution:

- Herbicides, pesticides and chemical fertilisers all get washed into rivers and streams.
- Fertilisers and untreated sewage can cause a high level of nitrates in the water.
- Toxic chemicals from landfill also leak into the waterways and pollute the water, killing organisms such as fish.

Key words: parasite, fertile

Student Book pages 268–269 **B3**

4.3 Air pollution

Key points

- Sulfur dioxide dissolved in rain makes it more acid.
- Acid rain can affect organisms both directly and indirectly.

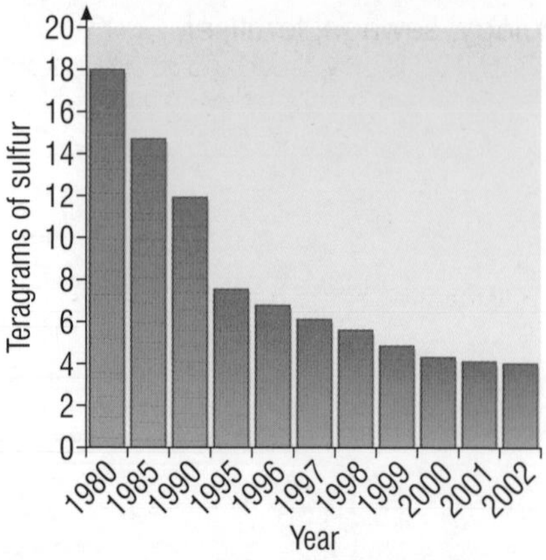

Graph to show the reductions in sulfur emissions made by European countries

- Burning fuels can produce sulfur dioxide and other acidic gases. Power stations and cars release acidic gases.
- The sulfur dioxide dissolves in water in the air, forming acidic solutions.
- The solutions then fall as acid rain – sometimes a long way from where the gases were produced.
- Acid rain kills organisms. Trees can be damaged if the leaves are soaked in acid rain for long periods.
- Acid rain can change the soil pH, which damages roots and may release toxic minerals. For example, aluminium ions are released which also damages organisms in the soil and in waterways.
- Enzymes, which control reactions, are very sensitive to pH (acidity or alkalinity).
- When trees are damaged, food and habitats for many other organisms are lost.

1 *How are birds affected by acid rain?*

Study tip

Remember that acid rain can affect areas a long way from where the sulfur dioxide is released, because the gas is blown by winds before it dissolves in rain.

4.4 Deforestation and peat destruction

Deforestation means that many trees are cut down. Large scale deforestation in tropical areas is due to the need for timber and to provide land for agriculture.

- Deforestation has:
  - increased the release of carbon dioxide into the atmosphere due to burning of the trees or decay of the wood by microorganisms,
  - reduced the rate at which carbon dioxide is removed from the atmosphere, by photosynthesis,
  - reduced biodiversity due to loss of food and habitats.

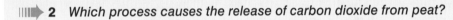 **1** *Which process removes carbon dioxide from the air?*

- Deforestation has occurred so that:
  - crops can be grown to produce ethanol-based biofuels,
  - there can be increases in cattle and rice fields for food.
- Cattle and rice growing produce **methane**, which has led to an increase of methane in the atmosphere.
- The destruction of peat bogs, and other areas of peat, also results in the release of carbon dioxide into the atmosphere. This occurs because the peat is removed from the bogs and used in compost for gardens. The compost is decayed by microorganisms.
- Using peat-free composts means the peat bogs will not be destroyed.

2 *Which process causes the release of carbon dioxide from peat?*

Key points

- Forests are cut down to clear land for faming.
- Deforestation leads to a reduction in biodiversity and an increase in carbon dioxide in the atmosphere.
- Destruction of peat bogs also releases carbon dioxide.

Bump up your grade

To get maximum marks in a question try to relate cause and effect. For example, make sure you understand why trees are important. Photosynthesis locks up carbon dioxide in plants. When trees are cut down and burned or decay, the carbon dioxide is released.

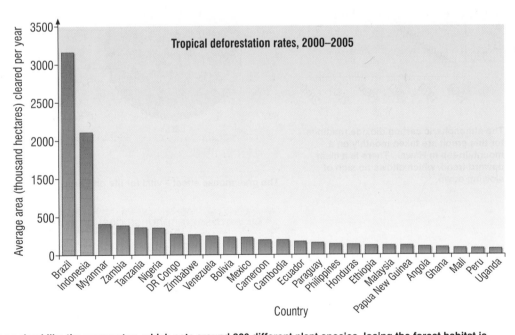

Tropical deforestation rates, 2000–2005

Average area (thousand hectares) cleared per year

Country

Brazil, Indonesia, Myanmar, Zambia, Tanzania, Nigeria, DR Congo, Zimbabwe, Venezuela, Bolivia, Mexico, Cameroon, Cambodia, Ecuador, Paraguay, Philippines, Honduras, Ethiopia, Malaysia, Papua New Guinea, Angola, Ghana, Mali, Peru, Uganda

The rate of deforestation is devastating. For an animal like the orang-utan, which eats around 300 different plant species, losing the forest habitat is driving the species to extinction.

Key words: deforestation, methane

Student Book
pages 272–273

B3

4.5 Global warming

- The climate of Earth is getting warmer.
- Global warming affects living organisms.

Study tip

Questions about the environment often include graphs and other data. Make sure you refer to the data in your answers.

- In the normal balance of nature, carbon dioxide is released into the air by respiration and removed by plants and algae in photosynthesis.
- Carbon dioxide also dissolves in oceans, rivers, lakes and ponds.
- We say that the carbon dioxide is **sequestered** by plants and water.

1 *How is carbon dioxide sequestered?*

Levels of carbon dioxide and methane are increasing in the atmosphere. They are called **greenhouse gases** and cause the **greenhouse effect**. Most scientists believe an increase in greenhouse gases contributes to **global warming**.

- An increase in the Earth's temperature of only a few degrees Celsius may:
  - Cause big changes in the Earth's climate.
  - Cause a rise in sea level due to melting of ice caps and glaciers.
  - Reduce biodiversity.
  - Cause changes in migration patterns e.g. of birds.
  - Result in changes in the distribution of species.

2 *Why might sea levels rise in the future?*

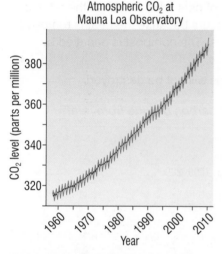

Atmospheric CO$_2$ at Mauna Loa Observatory

The atmospheric carbon dioxide readings for this graph are taken monthly on a mountain-top in Hawaii. There is a clear upward trend, which shows no sign of slowing down.

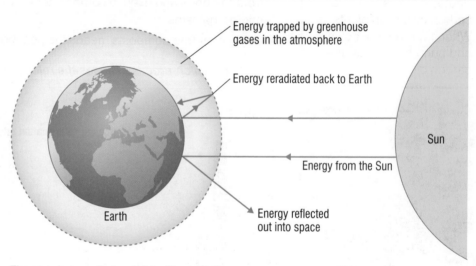

The greenhouse effect – vital for life on Earth

Key words: greenhouse gas, greenhouse effect, global warming

Student Book
pages 274–275

B3

4.6 Biofuels

Biofuels are made from natural products. Two types of biofuel are ethanol-based fuels and **biogas**.

- Ethanol-based fuels can be produced by fermentation.
- Microorganisms respire anaerobically to produce the ethanol, using sugars from crops as the energy source.
- Glucose is produced from maize starch by the action of a carbohydrase.
- The glucose and sugar cane juices can be fermented by yeast to produce ethanol.
- The ethanol is extracted by the process of **distillation** and can then be used as a fuel in motor vehicles.
- Using ethanol as a fuel could replace fossil fuels in the future. In terms of the 'greenhouse effect', using ethanol as a fuel is much more 'carbon friendly'.
- Ethanol is described as **carbon neutral** because only the carbon dioxide used for photosynthesis by the crops is returned to the atmosphere when the ethanol is burned.

IIIII➤ **1** *How is ethanol extracted from the fermenter?*

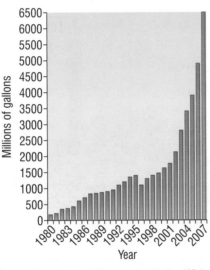

Increasing demand for gasohol in the US has led to increasing production of ethanol from maize, as this data clearly shows

Key words: biofuel, biogas, distillation, carbon neutral

Student Book
pages 276–277 **B3**

4.7 Biogas

This commercial biogas plant in Texas uses the slurry from 10 000 cows as well as other agricultural waste as its raw material

Study tip

Remember that biogas is produced by fermentation. The bacteria respire anaerobically using waste carbohydrate.

- Biogas, mainly methane, can be produced during anaerobic fermentation by bacteria.
- Plants and any waste material containing carbohydrate, e.g. cattle dung, can be broken down in biogas generators.
- The generators provide the ideal conditions for bacteria to reproduce and ferment the carbohydrates. They must be maintained at a suitable temperature in oxygen-free conditions. Some generators are designed to mix the contents. Generators are either buried in the ground for insulation or have insulating jackets.
- Large scale, commercial generators use waste from sugar factories or sewage works.
- On a small scale, generators can be used by a home or farm.
- The gas produced is a fuel and provides energy for heating etc. The more methane in the gas mixture, the better the quality of biogas.

1 *Which gas is the fuel in biogas?*

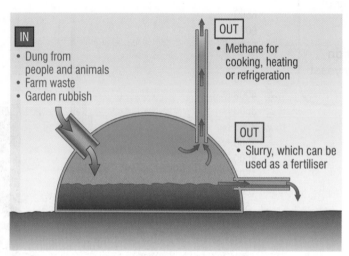

IN
- Dung from people and animals
- Farm waste
- Garden rubbish

OUT
- Methane for cooking, heating or refrigeration

OUT
- Slurry, which can be used as a fertiliser

Biogas generators take in body waste or plants, and biogas and useful fertilisers come out at the other end

Student Book
pages 278–279 **B3**

4.8 Making food production efficient

- The shorter the food chain, the less energy will be wasted. It is therefore more efficient for us to eat plants than it is to eat animals.
- We can produce meat more efficiently by:
  - preventing the animal from moving so it doesn't waste energy on movement; but this is seen as cruelty by many people and is controversial,
  - keeping the animal in warm sheds so it doesn't use as much energy from food to maintain its body temperature.

1 *Why are farm animals sometimes kept in warm sheds?*

Intensively reared chickens versus free range chickens

Student Book pages 280–281   **B3**

4.9 Sustainable food production

The human population is increasing rapidly and food resources are at risk of running out.

Sustainable food production involves managing resources, and finding new types of food such as mycoprotein. This ensures there is enough food for the current population and in the future.

- Fish stocks in the oceans are monitored.
- Fishermen:
  - can only remove a strict allocation of fish per year – a quota,
  - must use certain sized nets to avoid catching small, young fish.

> **1** *Why can't fishermen remove small, young fish?*

- The fungus *Fusarium* is grown to produce **mycoprotein**. This is a protein-rich food suitable for vegetarians. *Fusarium* is grown aerobically on cheap sugar syrup made from waste starch and the mycoprotein harvested.
- Microorganisms can be grown on a large scale in industrial **fermenters**.
- The conditions in a fermenter must be controlled to ensure maximum growth of the *Fusarium*.
- Industrial fermenters are large vessels which have:
  - an air supply providing oxygen for respiration,
  - a stirrer or gas bubbles used to keep the microorganisms spread out and to provide an even temperature,
  - a water-cooled jacket around the outside, as the respiring microorganisms release energy which heats the contents,
  - sensors to monitor both pH and temperature.

> **2** *What type of food is produced by Fusarium ?*

Key words: sustainable food production, mycoprotein

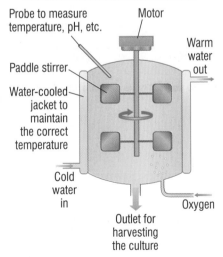

Probe to measure temperature, pH, etc.

Motor

Paddle stirrer

Warm water out

Water-cooled jacket to maintain the correct temperature

Cold water in

Oxygen

Outlet for harvesting the culture

Conditions inside the fermenters used to culture microorganisms such as Fusarium are kept as stable as possible

4.10 Environmental issues

Key points

- Human activities affect both local and global environments.
- There is a lot of evidence for environmental change.
- Scientists need to check the validity and the reproducibility of the data collected.

Bump up your grade

To improve your grade when answering questions on environmental issues, always read the evidence and data given in the stem of the question. You will be expected to quote from the data to back up your arguments.

- There are many human activities which can affect the global environment. These include:
  - deforestation which can cause increased levels of carbon dioxide in the atmosphere,
  - increases in rice growing and rearing cattle resulting in more methane being released,
  - building dams, to store water in reservoirs, causing loss of habitats, drying out of rivers below the dams and reduction in fertile land to grow crops.

▐▐▐▶ **1** *Why do humans build dams?*

- There are huge amounts of environmental data produced by many different scientists in many different countries. It can be very difficult to be sure the data are valid and reliable.
- Scientists often come to different conclusions even when considering the same data. Explanations sometimes depend on the individual opinions of the scientist and can be biased.
- The issue of global warming divides opinion. Many people think the Earth's temperature has increased due to increases in greenhouse gases. Others say the increase is part of a natural cycle.

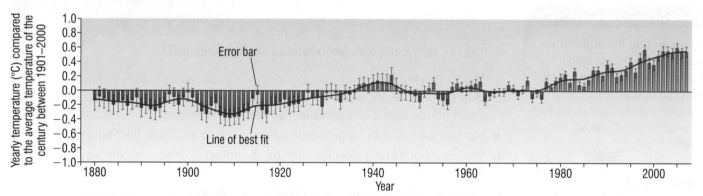

This graph shows how global surface temperatures have varied from the 1901–2000 mean over 130 years. These data are widely regarded as very reliable and valid.

1 What are herbicides?

2 Which gas causes acid rain?

3 Why does acid rain often fall a great distance from where it was produced?

4 How do soil fertilisers cause damage to rivers?

5 How is methane produced?

6 When trees are cut down, carbon dioxide is released into the atmosphere. Explain how.

7 Explain why using ethanol-based fuels can be described as being carbon neutral.

8 Why must biogas be produced anaerobically?

9 Give three factors which must be controlled in an industrial fermenter that produces mycoprotein.

10 Why do many scientists think that human activities are causing global warming?

Chapter checklist ✓✓✓

Tick when you have:

reviewed it after your lesson ☑ ☐ ☐

revised once – some questions right ☑ ☑ ☐

revised twice – all questions right ☑ ☑ ☑

Move on to another topic when you have all three ticks

The effects of the population explosion	☐	☐	☐
Land and water pollution	☐	☐	☐
Air pollution	☐	☐	☐
Deforestation and peat destruction	☐	☐	☐
Global warming	☐	☐	☐
Biofuels	☐	☐	☐
Biogas	☐	☐	☐
Making food production efficient	☐	☐	☐
Sustainable food production	☐	☐	☐
Environmental issues	☐	☐	☐

1 Plants have transport tissues. Match the tissue to its correct function from the list.

   a phloem *(1 mark)*

   b xylem *(1 mark)*

   Functions: transports gases into the leaves, transports sugar to storage organs, transports water up the stem

2 The heart pumps blood around the body. This causes blood to leave the heart at high pressure.

The graph shows blood pressure measurements for a person at rest. The blood pressure was measured in an artery and in a vein.

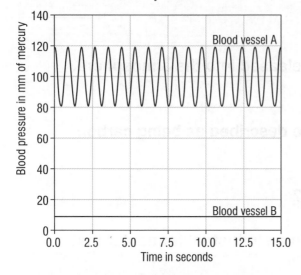

Study tip

Don't forget to read the axes' units. In Q2 the units of time are given in seconds, but you have to calculate the number of beats per minute.

   a Which blood vessel, **A** or **B**, is the artery?
Give **two** reasons for your answer. *(2 marks)*

   b Use information from the graph to answer these questions.

      i How many times did the heart beat in 15 seconds? *(1 mark)*

      ii Use your answer from part (b)(i) to calculate the person's heart rate in beats per minute. *(1 mark)*

   c During exercise, the heart rate increases. This supplies useful substances to the muscles and removes waste materials from the muscles at a faster rate.

      i Name **two** useful substances that must be supplied to the muscles at a faster rate during exercise. *(2 marks)*

      ii Name **one** waste substance that must be removed from the muscles at a faster rate during exercise. *(1 mark)*

AQA, 2009

3 The photograph shows an amoeba.

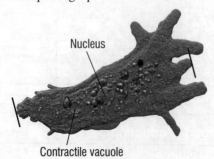

The amoeba is a single-celled organism with a cell membrane, nucleus and cytoplasm. The amoeba lives in fresh water. The cytoplasm is similar to most animal cells and also has various tiny vacuoles containing the remains of digested food.

a   The amoeba can change shape but assuming it is 0.5 mm long between the black lines, what is the magnification of the photograph?   *(2 marks)*

b   The contractile vacuole can fill with water and expel it from the amoeba.
  i   Explain why amoeba needs a contractile vacuole.   *(4 marks)*
  ii Some types of amoeba live in sea water. These amoebae do not have contractile vacuoles. Explain why.   *(2 marks)*

4   A fermenter is used for growing the fungus *Fusarium* which is used to make mycoprotein.

a   It is important to prevent microorganisms other than *Fusarium* from growing in the fermenter.
  i   Why is this important?   *(1 mark)*
  ii Suggest **two** ways in which contamination of the fermenter by microorganisms could be prevented.   *(2 marks)*

b   Human cells cannot make some of the amino acids which we need. We must obtain these amino acids from our diet. The table shows the amounts of four of these amino acids present in mycoprotein, in beef and in wheat.

Name of amino acid	Amount of amino acid per 100 g (mg)			Daily amount needed by a 70 kg human (mg)
	Mycoprotein	Beef	Wheat	
lysine	910	1600	300	840
methionine	230	500	220	910
phenylalanine	540	760	680	980
threonine	610	840	370	490

A diet book states that mycoprotein is the best source of amino acids for the human diet. Evaluate this statement. Remember to include a conclusion in your evaluation.   *(4 marks)*

AQA, 2008

5   The levels of carbon dioxide and methane in the atmosphere are increasing. Many scientists believe this increase may contribute to 'global warming'.

a   Describe **two** consequences of an increase of only a few degrees in the Earth's temperature.   *(2 marks)*

b   *In this question you will be assessed on using good English, organising information clearly and using specialist terms where appropriate.*

Describe the human activities which may have contributed to an increase in atmospheric carbon dioxide and methane.   *(6 marks)*

1.1 The early periodic table

Key points

- The periodic table of the elements developed as chemists tried to classify the elements. It arranges them in a pattern in which similar elements are grouped together.

- Newlands' table put the elements in order of atomic weight but failed to take account of elements that were unknown at that time.

- Mendeleev's periodic table left gaps for the unknown elements, and so provided the basis for the modern periodic table.

Dmitri Mendeleev together with a Russian stamp issued in his honour in 1969

- During the 19th century, many elements had been discovered but scientists did not know about the structure of atoms. Scientists tried to find ways to classify the elements based on their properties and **atomic weights**.

▐▐▶ **1** *Why did scientists in the 19th century use atomic weights rather than proton numbers to classify the elements?*

- In 1863 Newlands proposed his law of octaves, which stated that similar properties are repeated every eighth element. He put the 62 elements known at that time into seven groups according to their atomic weights. After calcium their properties did not match very well within the groups and so other scientists did not accept his ideas.

- In 1869 Mendeleev produced a better table. He left gaps for undiscovered elements so that the groups of known elements did have similar properties. He predicted the properties of the missing elements. When some of the missing elements were discovered, Mendeleev's predictions were confirmed and then other scientists more readily accepted his ideas. Mendeleev's table became the basis for the modern **periodic table**.

▐▐▶ **2** *Why was Mendeleev's table better than Newlands' table?*

> ### Study tip
>
> You may have studied the work of other scientists in developing the periodic table but you only need to know about the work of Newlands and Mendeleev for the exam. You do not need to remember details of the early tables.

				K = 39	Rb = 85	Cs = 133	—	—
				Ca = 40	Sr = 87	Ba = 137	—	—
				—	?Yt = 88?	?Di = 138?	Er = 178?	—
				Ti = 48?	Zr = 90	Ce = 140?	?La = 180?	Tb = 231
				V = 51	Nb = 94	—	Ta = 182	—
				Cr = 52	Mo = 96	—	W = 184	U = 240
				Mn = 55	—	—	—	—
				Fe = 56	Ru = 104	—	Os = 195?	—
Typische Elemente				Co = 59	Rh = 104	—	Ir = 197	—
				Ni = 59	Pd = 106	—	Pt = 198?	—
H = 1	Li = 7	Na = 23		Cu = 63	Ag = 108	—	Au = 199?	—
	Be = 9,4	Mg = 24		Zn = 65	Cd = 112	—	Hg = 200	—
	B = 11	Al = 27,3		—	In = 113	—	Tl = 204	—
	C = 12	Si = 28		—	Sn = 118	—	Pb = 207	—
	N = 14	P = 31		As = 75	Sb = 122	—	Bi = 208	—
	O = 16	S = 32		Se = 78	Te = 125?	—	—	—
	F = 19	Cl = 35,5		Br = 80	J = 127	—	—	—

An early version of Mendeleev's periodic table with similar elements arranged horizontally in rows

Key words: atomic weight, periodic table

C3

1.2 The modern periodic table

- Scientists found out about protons and electrons at the start of the 20th century. Soon after this, they developed models of the arrangement of electrons in atoms. The elements were arranged in the periodic table in order of their atomic numbers (proton numbers) and were lined up in vertical **groups**.

- The groups of elements have similar chemical properties because their atoms have the same number of electrons in their highest occupied energy level (outer shell). For the main groups, the number of electrons in the outer shell is the same as the group number.

> **1** *Why do elements in a group have similar chemical properties?*

Reactivity within groups

Within a group the reactivity of the elements depends on the total number of electrons. Going down a group, there are more occupied energy levels and the atoms get larger. As the atoms get larger, the electrons in the highest occupied energy level (outer shell) are less strongly attracted by the nucleus.

- When metals react they lose electrons, so the reactivity of metals in a group increases going down the group.

- When non-metals react they gain electrons, so the reactivity of non-metals decreases going down a group.

> **2** *Why do metals get more reactive going down a group?*

Key points

- The atomic (proton) number of an element determines its position in the periodic table.

- The number of electrons in the outermost shell (highest energy level) of an atom determines its chemical properties.

- The group number in the periodic table equals the number of electrons in the outermost shell.

- We can explain trends in reactivity as we go down a group in terms of:
  - the distance between the outermost electrons and the nucleus
  - the number of occupied inner shells (energy levels) in the atoms. **[H]**

Bump up your grade

You should be able to explain the trends in reactivity in the main groups in the periodic table in terms of electronic structure.

Study tip

You may describe electron arrangements in terms of energy levels or shells. The term 'outer electrons' is generally accepted as referring to the electrons in the highest occupied energy level or outer shell.

Higher

Group numbers																	0
1	2											3	4	5	6	7	4 He 2
7 Li 3	9 Be 4				Relative atomic mass ⟶ 1							11 B 5	12 C 6	14 N 7	16 O 8	19 F 9	20 Ne 10
23 Na 11	24 Mg 12				Atomic (proton) number ⟶ 1	H						27 Al 13	28 Si 14	31 P 15	32 S 16	35.5 Cl 17	40 Ar 18
39 K 19	40 Ca 20	45 Sc 21	48 Ti 22	51 V 23	52 Cr 24	55 Mn 25	56 Fe 26	59 Co 27	59 Ni 28	63.5 Cu 29	65 Zn 30	70 Ga 31	73 Ge 32	75 As 33	79 Se 34	80 Br 35	84 Kr 36
85 Rb 37	88 Sr 38	89 Y 39	91 Zr 40	93 Nb 41	96 Mo 42	98 Tc 43	101 Ru 44	103 Rh 45	106 Pd 46	108 Ag 47	112 Cd 48	115 In 49	119 Sn 50	122 Sb 51	128 Te 52	127 I 53	131 Xe 54
133 Cs 55	137 Ba 56	139 La 57	178 Hf 72	181 Ta 73	184 W 74	186 Re 75	190 Os 76	192 Ir 77	195 Pt 78	197 Au 79	201 Hg 80	204 Tl 81	207 Pb 82	209 Bi 83	209 Po 84	210 At 85	222 Rn 86
223 Fr 87	226 Ra 88	227 Ac 89															

Elements 58–71 and 90–103 (all metals) have been omitted

The modern periodic table

Student Book
pages 200–201 **C3**

1.3 Group 1 – the alkali metals

- The Group 1 elements are called the **alkali metals**. They are all metals that react readily with air and water.
- They are soft solids at room temperature with low melting and boiling points that decrease going down the group. They have low densities, so lithium, sodium and potassium float on water.
- They react with water to produce hydrogen gas and a metal hydroxide that is an alkali, e.g.

$$\text{sodium} + \text{water} \rightarrow \text{sodium hydroxide} + \text{hydrogen}$$
$$2Na(s) + 2H_2O(l) \rightarrow 2NaOH(aq) + H_2(g)$$

▶ **1** *Why are the elements in Group 1 called 'alkali metals'?*

- They all have one electron in their highest occupied energy level (outer shell). They lose this electron in reactions to form ionic compounds in which their ions have a single positive charge, e.g. Na^+.
- They react with the halogens (Group 7) to form salts that are white or colourless crystals, e.g.

$$\text{sodium} + \text{chlorine} \rightarrow \text{sodium chloride}$$
$$2Na(s) + Cl_2(g) \rightarrow 2NaCl(s)$$

- Compounds of alkali metals dissolve in water, forming solutions that are usually colourless.
- Going down Group 1, the reactivity of the alkali metals increases.

▶ **2** *Name and give the formula of the compound formed when potassium reacts with bromine.*

Explanation of reactivity trend in Group 1

Reactivity increases going down Group 1 because the outer electron is less strongly attracted to the nucleus as the number of occupied energy levels increases and the atoms get larger.

▶ **3** *Why is lithium less reactive than sodium?*

Bump up your grade

If you are taking the Higher Tier paper, you should be able to explain the trend in reactivity in Group 1 in terms of electronic structure.

Key word: alkali metal

Key points

- The elements in Group 1 of the periodic table are called the alkali metals.
- These metals all react with water to produce hydrogen and an alkaline solution containing the metal hydroxide.
- They form positive ions with a charge of 1+ in reactions to make ionic compounds. Their compounds are usually white or colourless crystals that dissolve in water producing colourless solutions.
- The reactivity of the alkali metals increases going down the group.

Study tip

The alkali metals form only ionic compounds in which their ions have a single positive charge.

| 7 Li 3 |
| 23 Na 11 |
| 39 K 19 |
| 85 Rb 37 |
| 113 Cs 55 |
| 223 Fr 87 |

The alkali metals (Group 1)

Student Book
pages 202–203

C3

1.4 The transition elements

- The **transition elements** are found in the periodic table between Groups 2 and 3.
- They are all metals and so are sometimes called the **transition metals**.

45 Sc 21	48 **Ti** 22	51 V 23	52 **Cr** 24	55 **Mn** 25	56 **Fe** 26	59 **Co** 27	59 **Ni** 28	63 **Cu** 29	64 **Zn** 30
89 Y 39	91 Zr 40	93 Nb 41	96 Mo 42	99 Tc 43	101 Ru 44	103 Rh 45	106 Pd 46	108 **Ag** 47	112 Cd 48
	178 Hf 72	181 Ta 73	184 W 74	186 Re 75	190 Os 76	192 Ir 77	195 **Pt** 78	197 **Au** 79	201 **Hg** 80

The transition elements. The more common elements are shown in bold type.

- Except for mercury, they have higher melting and boiling points than the alkali metals.
- They are malleable and ductile and they are good conductors of heat and electricity.
- They react only slowly, or not at all, with oxygen and water at ordinary temperatures.
- Most are strong and dense and are useful as building materials, often as alloys.

▷ **1** *Why are transition metals useful as building materials?*

- They form positive ions with various charges, e.g. Fe^{2+} and Fe^{3+}.
- Compounds of transition metals are often brightly coloured.
- Many transition metals and their compounds are catalysts for chemical reactions.

▷ **2** *List the ways in which transition elements are different from the elements in Group 1.*

Transition metals are used as building materials, e.g. iron is used in the steel in this bridge

Key words: transition element, transition metals

Key points

- Compared with the alkali metals, transition elements have much higher melting points and densities. They are also stronger and harder, but are much less reactive.
- The transition elements do not react vigorously with oxygen or water.
- Transition elements can form ions with different charges, in compounds that are often coloured.
- Transition elements and their compounds are important industrial catalysts.

Bump up your grade

For your exam, try to remember how to write formulae for transition metal compounds. If you are taking the Higher Tier paper you should also be able to balance equations for the reactions of transition metals.

Study tip

The charge on a transition metal ion is given by the Roman numeral in its name. For example, iron(II) chloride contains Fe^{2+} ions and so its formula is $FeCl_2$ and iron(III) chloride contains Fe^{3+} ions and its formula is $FeCl_3$.

Student Book
pages 204–205

C3

1.5 Group 7 – the halogens

- The **halogens** are non-metallic elements in Group 7 of the periodic table.
- They exist as small molecules made up of pairs of atoms. They have low melting and boiling points that increase going down the group. At room temperature fluorine is a pale yellow gas, chlorine is a green gas, bromine is a red-brown liquid and iodine is a grey solid. Iodine easily vaporises to a violet gas.

> **1** Why do the halogens have low melting and boiling points?

- All of the halogens have seven electrons in their highest occupied energy level.
- The halogens form ionic compounds with metals in which the **halide ions** have a charge of 1–.
- The halogens also bond covalently with non-metals, forming molecules.
- The reactivity of the halogens decreases going down the group. A more reactive halogen is able to displace a less reactive halogen from an aqueous solution of a halide compound.

> **2** How could you show that chlorine is more reactive than bromine?

Key points

- The halogens all form ions with a single negative charge in their ionic compounds with metals.
- The halogens form covalent compounds by sharing electrons with other non-metals.
- A more reactive halogen can displace a less reactive halogen from a solution of one of its salts.
- The reactivity of the halogens decreases going down the group.

19	
F	
9	
35	
Cl	
17	
80	
Br	
35	
127	
I	
53	
210	
At	
85	

The Group 7 elements, the halogens

Explanation of reactivity trend for Group 7

The reactivity of the halogens decreases going down Group 7 because the attraction of the outer electrons to the nucleus decreases as the number of occupied energy levels (shells) increases.

Study tip

Make sure you revise ionic and covalent bonding so you are clear about the differences in properties between ionic compounds and covalent compounds that have small molecules.

Bump up your grade

If you are taking the Higher Tier paper, you should be able to explain the trend in reactivity in Group 7 in terms of electronic structure.

Chlorine, bromine and iodine

1 What was Newlands' law of octaves?

2 How was Mendeleev's periodic table an improvement on Newlands' table?

3 Why do elements in the groups in the modern periodic table have similar properties?

4 A small piece of lithium is added to a bowl of water.
  a Write a word equation for the reaction of lithium with water.
  b Describe three things that you would see when the lithium is added to the water.
  c How could you show that an alkali is produced?
  d Give one way in which the reaction of sodium with water is different to the reaction of lithium with water.

5 Predict three physical and three chemical properties of the transition element cobalt, Co.

6 What is the trend in melting points and boiling points going down Group 7?

7 What is the formula of sodium bromide? Describe its appearance and what happens when it is mixed with water.

8 Hydrogen chloride is a gas at room temperature. Explain why.

9 Some chlorine water was added to an aqueous solution of potassium bromide.
  a Describe the colour change that you would see.
  b Write a word equation for the reaction that happens.
  c Write a balanced symbol equation for the reaction.   [H]

10 Iron reacts with chlorine to produce iron(III) chloride. Write a balanced symbol equation for this reaction.   [H]

11 Explain in terms of electronic structures:
  a why sodium is more reactive than lithium
  b why fluorine is more reactive than chlorine.   [H]

Chapter checklist	✔ ✔ ✔

Tick when you have:		The early periodic table	☐ ☐ ☐
reviewed it after your lesson	✔ ☐ ☐	The modern periodic table	☐ ☐ ☐
revised once – some questions right	✔ ✔ ☐	Group 1 – the alkali metals	☐ ☐ ☐
revised twice – all questions right	✔ ✔ ✔	The transition elements	☐ ☐ ☐
Move on to another topic when you have all three ticks		Group 7 – the halogens	☐ ☐ ☐

2.1 Hard water

- Hard water contains dissolved compounds such as calcium and magnesium salts.

- The calcium and/or magnesium ions in hard water react with soap producing a precipitate called scum.

- Temporary hard water can produce a solid scale when it is heated, reducing the efficiency of heating systems and kettles.

- Hard water is better than soft water for developing and maintaining teeth and bones. It may also help to prevent heart disease.

- Water that lathers easily with soap is said to be **soft water**. **Hard water** uses more soap to produce lather and to wash effectively. This is because hard water contains dissolved compounds that react with soap to form an insoluble solid called **scum**. Other detergents, called **soapless detergents**, do not react with hard water to form scum.

▶ **1** *What is the difference between soft water and hard water?*

- When water is in contact with rocks some compounds dissolve. If the water contains dissolved calcium or magnesium ions, these will react with soap to form scum and so the water is hard.

- When it is heated, one type of hard water, called temporary hard water, produces an insoluble solid called **scale**. Scale can be deposited in kettles, boilers and pipes. This reduces the efficiency of heating systems and causes blockages.

▶ **2** *What is the difference between scum and scale?*

- Calcium compounds are good for our health, helping to develop strong bones and teeth. Calcium may also reduce the risk of heart disease.

▶ **3** *Why is it better to drink hard water rather than soft water?*

> **Study tip**
>
> Many candidates confuse scum and scale. **S**cu**m** is formed when **s**oap reacts with dissolved co**m**pounds in hard water. When temporary hard water is heated it produces **scale** which **covers** pipes and heating elements (**scales cover fish**).

Scum is left in the sink after using soap with hard water

As scale builds up in heating systems and kettles it not only makes them less efficient – it can stop them working completely

Key words: soft water, hard water, scum, soapless detergent, scale

2.2 Removing hardness

Key points

- Soft water does not contain salts that produce scum or scale.

- Hard water can be softened by removing the salts that produce scum and scale.

- Temporary hardness is removed from water by heating it. Permanent hardness is not changed by heating.

- The hydrogencarbonate ions in temporary hard water decompose on heating. The carbonate ions formed react with Ca^{2+}(aq) and Mg^{2+}(aq), making precipitates. **[H]**

- Both types of hard water can be softened by adding washing soda or by using an ion-exchange resin to remove calcium and magnesium ions.

- Soft water may contain dissolved substances but it does not contain dissolved salts that react with soap to produce scum. Also, it does not produce scale when it is heated.

- Hard water can be made soft by removing the dissolved calcium and magnesium ions that react with soap.

- Some types of hard water are affected by heating while others are not. **Temporary hard water** is softened by boiling because when it is heated the calcium and magnesium compounds form insoluble scale and this removes them from the water. **Permanent hard water** is not softened by boiling and does not produce scale when it is heated.

> **1** What is meant by 'temporary hard water'?

How temporary hard water is softened by heating

Temporary hard water contains hydrogencarbonate ions, HCO_3^-(aq). The hydrogencarbonate ions decompose when heated to produce carbonate ions, water and carbon dioxide:

$$2HCO_3^-(aq) \rightarrow CO_3^{2-}(aq) + H_2O(l) + CO_2(g)$$

The carbonate ions react with calcium ions and magnesium ions in the water to produce precipitates of calcium carbonate and magnesium carbonate that are deposited as scale.

> **2** Write a balanced equation, including state symbols, for the reaction of calcium ions with carbonate ions.

Study tip

Any method that removes dissolved calcium and magnesium ions from hard water will soften the water.

- One method of softening either type of hard water is by precipitating out the ions that cause hardness. This can be done by adding washing soda, which is sodium carbonate. The sodium carbonate reacts with calcium ions and magnesium ions in the water to form solid calcium carbonate and magnesium carbonate that cannot react with soap.

- Another method is to use an **ion-exchange column** packed with a resin containing sodium or hydrogen ions. When hard water is passed through the resin, calcium and magnesium ions become attached to the resin and sodium ions or hydrogen ions take their place in the water. Sodium ions and hydrogen ions do not react with soap.

> **3** How does an ion-exchange resin soften hard water?

Bump up your grade

If you are taking the Higher Tier paper, you should be able to explain and write equations for the reactions that happen when temporary hard water is heated.

Key words: temporary hard water, permanent hard water, ion-exchange column

Washing soda is a simple way to soften water

Student Book
pages 212–213 **C3**

2.3 Water treatment

- Drinking water should not contain any harmful substances and should have sufficiently low levels of dissolved salts and microbes.
- Water from an appropriate source can be treated to make it safe to drink. Water is often treated by sedimentation and filtration to remove solids. This is followed by disinfection to kill microbes in the water. Chlorine is often used to kill microbes in drinking water.

> **1** *How is drinking water treated to make it safe to drink?*

- Water filters can be used to improve the taste of water. They often contain carbon and an ion-exchange resin that remove some soluble substances and silver or another substance to prevent the growth of bacteria.
- Pure water can be made by distillation. This requires a large amount of energy to boil the water and so it would be very expensive to do on a large scale.

> **2** *Explain why distillation is not used to treat mains tap water.*

Key points

- Water for drinking should contain only low levels of dissolved substances and microbes.
- Water is made fit to drink by filtering it to remove solids and adding chlorine to kill microbes.
- We can make pure water by distillation but this requires large amounts of energy which makes it expensive.

Good, clean water is a precious resource. Those of us lucky enough to have it can too easily take it for granted.

Study tip

Many people think that drinking water must be pure water. It does not have to be pure, but it should not contain anything that will cause us harm.

Bump up your grade

You should know the three main stages involved in producing water that is fit to drink: suitable source, removal of solids, killing of microbes.

Student Book
pages 214–215 **C3**

2.4 Water issues

- When water is treated, there are advantages and disadvantages. These must be carefully considered before any decision to treat water is taken. This is particularly important for the treatment of public water supplies.
- The hardness of the water supplied depends on where you live. Hard water causes problems in heating systems and with washing, but if used for drinking has health benefits. If the water is not suitable for a particular purpose you can treat the water or use an alternative supply.

> **1** *Suggest one advantage and one disadvantage of softening hard water.*

- Chlorine is particularly effective in killing microbes in water so that it is safe to use. However, chlorine is poisonous, and it can produce other toxic compounds. Therefore its use must be carefully controlled to minimise the risks.
- Fluoride compounds are added to toothpastes and to water supplies to help prevent tooth decay. The arguments for and against adding fluorides to water supplies are complicated. One of the arguments against adding fluoride to water is that people should be able to choose to take extra fluoride or not.

> **2** *What would be the consequences of not adding chlorine and fluoride to tap water?*

Key points

- There are advantages and disadvantages to any type of water treatment.
- Water can be treated to remove hardness, to remove harmful microbes and to improve dental health.

Study tip

You should understand the principles of water treatment described in the specification but you do not need to remember any specific details of other methods. However, you should be prepared to evaluate any information you are given about water treatment methods.

1 Why does hard water produce scum?

2 What is produced when temporary hard water is heated?

3 Give one advantage of drinking hard water.

4 Explain the difference between temporary hard water and permanent hard water.

5 How does washing soda soften hard water?

6 Give one disadvantage of using a sodium ion-exchange resin to soften hard water.

7 What type of substance is removed from water at a treatment works by filtration?

8 Why is chlorine used in water treatment?

9 Why is it not necessary to distil water used for drinking?

10 Why are fluorides added to drinking water?

11 Explain what happens when temporary hard water is boiled. Include two balanced symbol equations in your answer. [H]

Chapter checklist			✓ ✓ ✓
Tick when you have:			Hard water
reviewed it after your lesson	✓ ☐ ☐		Removing hardness
revised once – some questions right	✓ ✓ ☐		Water treatment
revised twice – all questions right	✓ ✓ ✓		Water issues
Move on to another topic when you have all three ticks			

Student Book
pages 218–219

C3

3.1 Comparing the energy released by fuels

Key points

- When fuels and food react with oxygen, energy is released in an exothermic reaction.

- A simple calorimeter can be used to compare the energy released by different fuels or different foods in a school lab.

Bump up your grade

Remember that you can calculate the energy released by burning a fuel using the equation $Q = mc\Delta T$. If you are taking the Higher Tier paper, you should be able to calculate the energy released by burning a known mass of fuel in kJ/g and, when given the formula of the fuel or its relative formula mass, in kJ/mol.

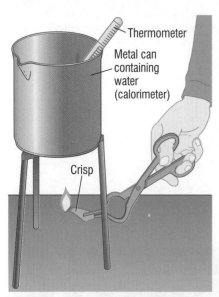

The energy released by fuels and foods when they burn can be compared using some very simple equipment

- When fuels and foods react with oxygen the reactions are exothermic. Different amounts of energy are released by different fuels and foods. The amount of energy released is usually measured in joules (J), but sometimes values are given in calories (1 cal = 4.2 J).

- We can use a calorimeter to measure the amount of energy released when substances burn. The simplest calorimeter is some water in a glass beaker or metal can. When a substance burns and heats the water, the temperature rise of the water depends on the amount of energy released.

- The amount of energy transferred to the water can be calculated using the equation:

$$Q = mc\Delta T$$

Where:

Q is the amount of energy transferred to the water in joules, J
m is the mass of water in grams, g
c is the specific heat capacity of water in J/g°C
ΔT is the temperature change in °C.

 Maths skills

Worked example

0.50 g of a fuel was burned and used to heat 200 g of water in a calorimeter. The temperature of the water increased by 14 °C. The specific heat capacity of water, $c = 4.2$ J/g °C

Using $Q = mc\Delta T$ Energy released = 200 × 4.2 × 14 = 11 760 J = 11.76 kJ
Energy released per g of fuel = 11.76/0.5 = 23.52 kJ/g

> **1** 0.45 g of fuel A was burned and heated 150 g of water in a calorimeter. The temperature of the water changed from 19 °C to 45 °C. How much energy was released by 1.0 g of fuel?

- Simple calorimeters do not give accurate results for the energy released because much of it is used to heat the surroundings. However, the results can be used to compare the energy released by different fuels.

- To compare the energy released by burning different substances either the energy change in kJ per gram or the energy change in kJ per mole can be used.

- The energy change in kJ/mol can be calculated by multiplying the energy change in kJ/g by the relative formula mass of the substance.

> **2** In a similar experiment to that in Question 1, fuel B released 35.6 kJ/g. The relative formula mass of fuel A is 72 and fuel B is 114. Which fuel releases more energy per mole?

Study tip

You do not need to remember the specific heat capacity of water or the value used to convert joules into calories because these will be given in any question when you need to use them.

C3

3.2 Energy transfers in solutions

Key points

- We can calculate the energy change for reactions in solution by measuring the temperature change and using the equation:

  $Q = mc\Delta T$

- Neutralisation and displacement reactions are both examples of reactions that we can use this technique for.

Study tip

For reactions in aqueous solutions, remember to use only the volume of the solution when calculating the energy change and assume that the specific heat capacity of any aqueous solution is the same as water.

- When a reaction takes place in solution, energy is transferred to or from the solution.
- We can do the reactions in an insulated container to reduce energy transfers to the surroundings.
- We can measure the temperature change of the solution and use this to calculate the energy change using the equation $Q = mc\Delta T$.
- In these calculations we assume the solutions behave like water. This means that 1 cm³ of solution has a mass of 1 g and the specific heat capacity of the solution is 4.2 J/g °C.

Maths skills

Worked example

A student added 25 cm³ of dilute nitric acid to 25 cm³ of potassium hydroxide solution in a polystyrene cup. He recorded a temperature rise of 12 °C. Calculate the energy change.

$Q = mc\Delta T$ Volume of solution = 25 + 25 = 50 cm³

Energy change = 50 × 4.2 × 12
          = 2520 J = 2.52 kJ

▸ **1** *When 50 cm³ of sulfuric acid was added to 100 cm³ of sodium hydroxide in a polystyrene cup the temperature increased by 12 °C. Calculate the energy change.*

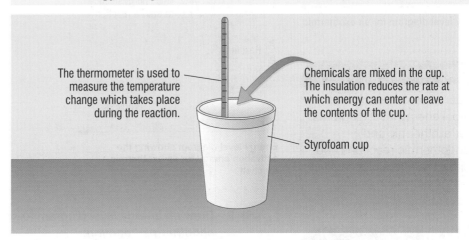

The thermometer is used to measure the temperature change which takes place during the reaction.

Chemicals are mixed in the cup. The insulation reduces the rate at which energy can enter or leave the contents of the cup.

Styrofoam cup

A simple calorimeter can be used to measure energy changes in solution

- When a solid is added to water or to an aqueous solution we assume that the volume of the solution does not change. We also assume that 1 cm³ of solution has a mass of 1 g and that its specific heat capacity is 4.2 J/g °C.
- If we know the number of moles involved in the reaction for which we have calculated the energy change we can calculate the energy change for the reaction in kJ/mol.

▸ **2** *When 5.6 g of iron filings reacted completely with 200 cm³ of copper(II) sulfate solution the temperature of the solution increased by 17 °C. Calculate the energy change in kJ/mol of iron. (Relative atomic mass of Fe = 56)*

3.3 Energy level diagrams

- We can show the relative difference in the energy of reactants and products on energy level diagrams.

- Catalysts provide a pathway with a lower activation energy so the rate of reaction increases.

- Bond breaking is endothermic and bond making is exothermic.

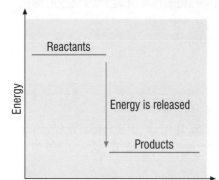

An energy level diagram for an exothermic reaction

Bump up your grade

Learn how to sketch and label energy level diagrams for exothermic and endothermic reactions and show the effect of a catalyst on the activation energy.

Study tip

Remember that energy increases up the diagrams, so endothermic changes (energy in) go up and exothermic changes (energy exits) go down.

- We can show the energy changes for chemical reactions on energy level diagrams.

- The difference between the energy levels of reactants and products is the energy change for the reaction.

- The energy level diagram for an exothermic reaction is shown on the left.

1 *Draw a similar energy level diagram for an endothermic reaction.*

- During a chemical reaction bonds in the reactants must be broken for the reaction to happen. Breaking bonds is endothermic because energy is taken in.

- The minimum energy needed for the reaction to happen is called the **activation energy**.

- When new bonds in the products are formed, energy is released and so this is exothermic.

- We can show the activation energy and how the energy changes during a reaction on an energy level diagram. This type of diagram is shown below.

2 *Draw a similar energy level diagram for an endothermic reaction.*

- Catalysts increase the rate of a reaction by providing a different pathway with an activation energy that is lower. The effect of a catalyst on an exothermic reaction is shown below.

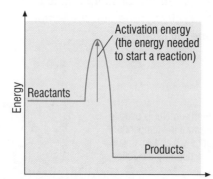

Energy level diagram showing the activation energy for an exothermic reaction

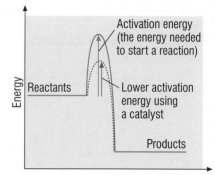

A catalyst provides a different reaction pathway with a lower activation energy

3 *Draw an energy level diagram to show the effect of a catalyst on an endothermic reaction.*

Key word: activation energy

Student Book pages 224–225 — C3

3.4 Calculations using bond energies

Key points

- In an exothermic reaction, the energy released when new bonds are formed is greater than the energy absorbed when bonds are broken. **[H]**

- In an endothermic reaction, the energy released when new bonds are formed is less than the energy absorbed when bonds are broken. **[H]**

- We can calculate the overall energy change in a chemical reaction using bond energies. **[H]**

Table of bond energies

Bond	Bond energy in kJ/mol
C–C	347
C–H	413
H–O	464
O=O	498
C=O	805

Bump up your grade

Get plenty of practice at calculating the energy change for a reaction given its balanced equation and values for the bond energies.

Bond energies

In a chemical reaction, energy is needed to break the bonds in the reactants. Energy is released when new bonds are formed in the products. It is the difference in these energy changes that makes the overall reaction exothermic or endothermic.

The energy needed to break the bond between two atoms is called the **bond energy** for that bond. An equal amount of energy is released when the bond forms between two atoms and so we can use bond energies to calculate the overall energy change for a reaction. Bond energies are measured in kJ/mol.

The balanced equation for the reaction is needed to calculate the energy change for a reaction. Then calculate:

- the total amount of energy needed to break all of the bonds in the reactants
- the total amount of energy released in making all of the bonds in the products
- the difference between the two totals.

Maths skills

Worked example

Use the bond energies in the table to calculate the energy change for burning methane:

$$CH_4 + 2O_2 \rightarrow CO_2 + 2H_2O$$

Bonds broken: $4 \times C–H + 2 \times O=O$ Energy needed $= (4 \times 413) + (2 \times 498)$
$$= 2648\,kJ$$

Bonds formed: $2 \times C=O + 4 \times H–O$ Energy released $= (2 \times 805) + (4 \times 464)$
$$= 3466\,kJ$$

Difference $= 3466 – 2648 = 818\,kJ$
Energy change for the reaction $= 818\,kJ/mol$ released

▶ 1 *Calculate the energy change for burning propane:*
$$C_3H_8 + 5O_2 \rightarrow 3CO_2 + 4H_2O$$

Study tip

Bond energies can be found in tables of data. In the examinations you will be given the values for any bond energies that you need to use in the questions.

You do not need to know about ΔH or sign conventions for the overall change – but you should be able to decide if the energy change calculated is energy released or energy absorbed (taken in).

Key word: bond energy

3.5 Fuel issues

Key points

- Much of the world relies on fossil fuels. However, they are non-renewable and they cause pollution. Alternative fuels need to be found soon.

- Hydrogen is one alternative. It can be burned in combustion engines or used in fuel cells to power vehicles.

Study tip

You should be able to link the ideas in this section to other parts of the specification. You should be prepared to consider and evaluate the use of fuels from your own knowledge and information you are given.

- Fossil fuels are non-renewable and they cause pollution. The need to develop alternative fuels is becoming more urgent.

- Hydrogen has advantages as a fuel. It burns easily and releases a large amount of energy per gram. It produces no carbon dioxide when it is burned, only water.

- Hydrogen can be burned in combustion engines or can be used in fuel cells to power vehicles.

- Hydrogen can be produced from renewable sources. The disadvantages of using hydrogen include supply, storage and safety problems.

- Vehicles that use fuel cells need to match the performance, convenience and costs of petrol and diesel vehicles.

1 *Why are cars fuelled with hydrogen being developed?*

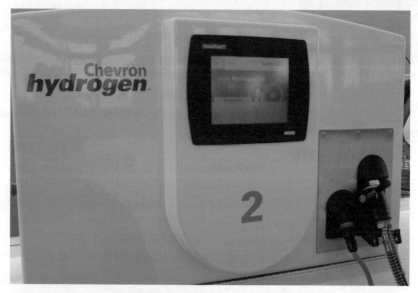

Some hydrogen re-fuelling stations have been set up to trial the use of hydrogen-powered combustion engines in vehicles

A London bus that runs on fuel cells

1 Why do simple calorimeters give inaccurate results for the energy released by burning a fuel?

2 How can you tell from an energy level diagram that a reaction is endothermic?

3 Draw an energy level diagram for the exothermic reaction $CH_4 + 2O_2 \rightarrow CO_2 + 2H_2O$. Show and label the reaction pathway, the activation energy and the energy change of the reaction.

4 0.50 g of a fuel was burned and heated 200 g of water in a calorimeter. The temperature increased by 15 °C. Calculate the energy released by the fuel using the equation $Q = mc\Delta T$, in which $c = 4.2$ J/g °C

5 0.10 mol of zinc was added to 100 cm³ of copper(II) sulfate solution. The temperature increased by 18 °C. Use the equation $Q = mc\Delta T$ to calculate the energy change for the reaction in kJ/mol.

6 100 cm³ of hydrochloric acid containing 0.10 mol of HCl was added to 100 cm³ of sodium hydroxide solution containing 0.10 mol of NaOH. The temperature of the solution increased by 7 °C. Calculate the energy change for this reaction in kJ/mol.

7 Draw and label an energy level diagram with reaction pathways to show the effect of a catalyst on an exothermic reaction.

8 Suggest three problems that need to be overcome to make hydrogen a suitable fuel for cars.

9 Calculate the energy change for the reaction $H_2 + F_2 \rightarrow 2HF$ using bond energies: H–H = 436 kJ/mol, F–F = 158 kJ/mol, H–F = 568 kJ/mol.   [H]

10 Calculate the energy change for the reaction $CH_2=CH_2 + 3O_2 \rightarrow 2CO_2 + 2H_2O$ using bond energies: C–H = 413 kJ/mol , C=C = 612 kJ/mol, O=O = 498 kJ/mol, C=O = 805 kJ/mol, H–O = 464 kJ/mol.   [H]

4.1 Tests for positive ions

- Some positive ions can be identified using a flame test or by using sodium hydroxide solution.
- Some metal ions produce colours when put into a flame:

Metal ion	Flame colour
Lithium (Li^+)	crimson (red)
Sodium (Na^+)	yellow
Potassium (K^+)	lilac
Calcium (Ca^{2+})	red
Barium (Ba^{2+})	green

 1 *Which metal ions give red colours in a flame?*

- The hydroxides of most metals that have ions with 2+ and 3+ charges are insoluble in water. When sodium hydroxide is added to solutions of these ions a precipitate of the metal hydroxide forms.
- Aluminium, calcium and magnesium ions form white precipitates. When excess sodium hydroxide solution is added the precipitate of aluminium hydroxide dissolves.
- Copper(II) hydroxide is blue.
- Iron(II) hydroxide is green.
- Iron(III) hydroxide is brown.

 2 *A few drops of sodium hydroxide solution were added to a colourless solution and a white precipitate appeared. When excess sodium hydroxide was added the precipitate remained. Which metal ions could be present?*

Equations for the reactions of positive ions with sodium hydroxide solution

We can show the reactions of positive ions with sodium hydroxide solution by balanced ionic equations. For example:

$$Fe^{3+}(aq) + 3OH^-(aq) \rightarrow Fe(OH)_3(s)$$

Bump up your grade

If you are taking the Higher Tier paper, you should be able to write balanced symbol equations for the reactions of positive ions with sodium hydroxide solution.

Key points

- Most Group 1 and Group 2 metal ions can be identified using flame tests.
- Sodium hydroxide solution can be used to identify different metal ions, depending on the precipitate that is formed.

Study tip

Both lithium ions and calcium ions give red flame colours. Lithium ions give a brighter red but it is difficult to tell them apart from this single test. Testing solutions of the ions with sodium hydroxide solution will show which is which because calcium ions will give a white precipitate but lithium ions will not.

Flame test colour of lithium ions

Student Book pages 232–233 **C3**

4.2 Tests for negative ions

Key points

- We identify carbonates by adding dilute acid, which produces carbon dioxide gas. The gas turns limewater cloudy.
- We identify halides by adding nitric acid, then silver nitrate solution. This produces a precipitate of silver halide (chloride = white, bromide = cream, iodide = pale yellow).
- We identify sulfates by adding hydrochloric acid, then barium chloride solution. This produces a white precipitate of barium sulfate.

Bump up your grade

If you are taking the Higher Tier paper, you should be able to write balanced symbol equations for the reactions in these tests.

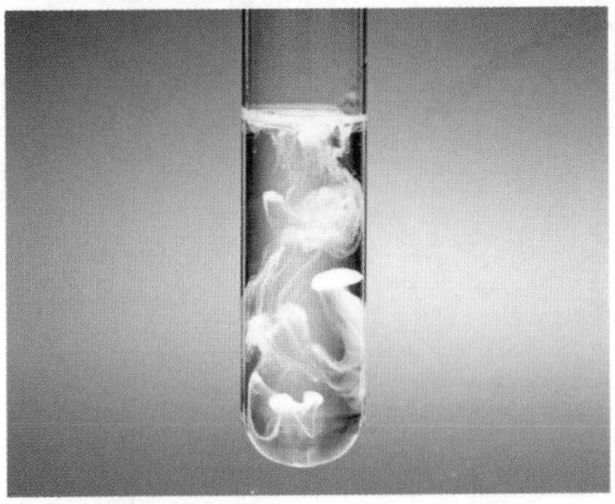

The white precipitate of barium sulfate

There are three tests for negative ions that you need to know.

- Carbonate ions: Add dilute hydrochloric acid to the substance to see if it fizzes. If it does and the gas produced turns limewater milky, the substance contains carbonate ions.

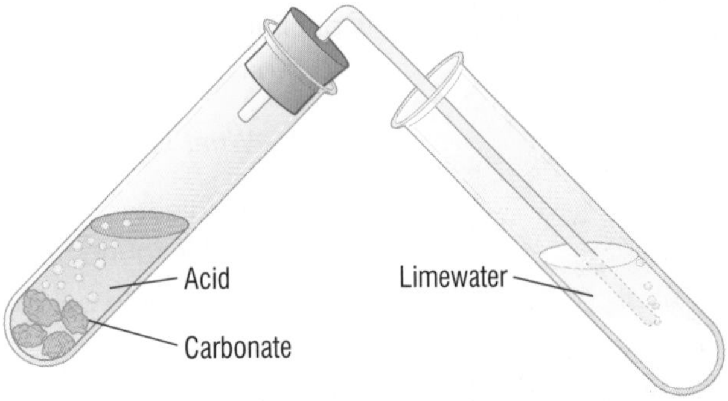

The test for carbonates

Higher

E.g. $2HCl(aq) + CaCO_3(s) \rightarrow CaCl_2(aq) + H_2O(l) + CO_2(g)$

- Halide ions: Add dilute nitric acid and then silver nitrate solution:
  - chloride ions give a white precipitate
  - bromide ions give a cream precipitate
  - iodide ions give a yellow precipitate.

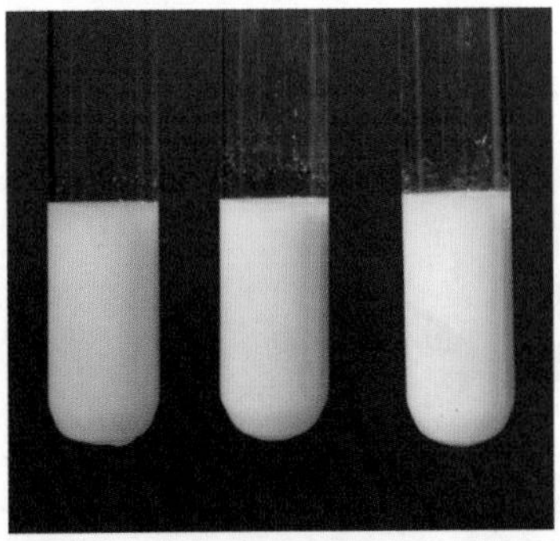

Precipitates of silver chloride, silver bromide and silver iodide

Higher

E.g. $AgNO_3(aq) + NaCl(aq) \rightarrow AgCl(s) + NaNO_3(aq)$

- Sulfate ions: Add dilute hydrochloric acid and then barium chloride solution. If a white precipitate forms, sulfate ions are present.

Higher

$BaCl_2(aq) + MgSO_4(aq) \rightarrow BaSO_4(s) + MgCl_2(aq)$

1 *Why must you add nitric acid and not hydrochloric acid or sulfuric acid when testing with silver nitrate solution for halides?*

Study tip

Make sure you learn these tests – many candidates lose marks because they do not know the tests or their results.

Student Book
pages 234–235 **C3**

4.3 Titrations

Key points

- A titration is used to measure accurately how much acid and alkali react together completely.

- The point at which an acid–base reaction is complete is called the end point of the reaction.

- We use an indicator to show the end point of the reaction between an acid and an alkali.

Study tip

Make sure you can describe how to use a pipette and a burette to do a titration and obtain results that are precise and repeatable.

- When solutions of an acid and an alkali react to form a salt and water, a neutralisation reaction takes place. The volumes of solutions that react exactly can be found by using a **titration**.

- To do a titration, a **pipette** is used to measure accurately the volume of alkali that is put into a conical flask. An indicator is added to the alkali. A **burette** is filled with acid, which is then added gradually to the flask.

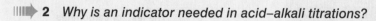 **1** *What is the difference between a pipette and a burette?*

- When the indicator changes colour the **end point** has been reached. The volume of acid used is found from the initial and final burette readings.

- The titration should be done several times to improve the repeatability of the results.

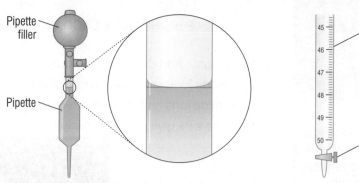 **2** *Why is an indicator needed in acid–alkali titrations?*

Pipette filler

Pipette

Burette

Tap

A pipette with a pipette filler attached and a burette

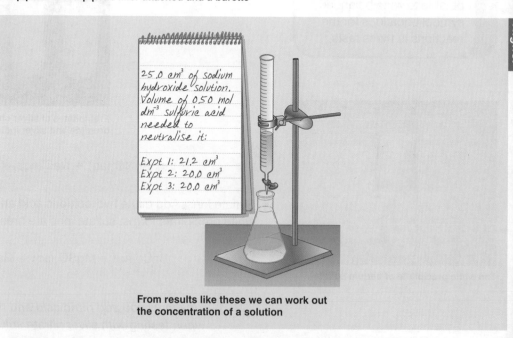

25.0 cm³ of sodium hydroxide solution.
Volume of 0.50 mol dm⁻³ sulfuric acid needed to neutralise it:

Expt 1: 21.2 cm³
Expt 2: 20.0 cm³
Expt 3: 20.0 cm³

From results like these we can work out the concentration of a solution

Titration experiment

Key words: titration, pipette, burette, end point

C3

4.4 Titration calculations

(handwritten: g/dm³ = (M/Σ) × 1000   md/dm³ = g/dm³ / R form mass)

(Higher)

Concentrations and titration calculations

Key points

- Concentrations of solutions can be measured in g/dm³ or mol/dm³.

- Concentrations can be calculated from the mass of solute dissolved in a known volume of solution.

- The mass of solute in any volume of solution can be calculated from its concentration.

- If the concentration of one of the solutions used in a titration is known, the results of the titration can be used to calculate the concentration of the other solution.

- Concentrations of solutions are measured in grams per decimetre cubed (g/dm³) or moles per decimetre cubed (mol/dm³).

- If we know the mass or the number of moles of a substance dissolved in a given volume of solution we can calculate its concentration.

- If we know the volume of a solution and its concentration we can calculate the mass or the number of moles of the substance in any volume of solution.

Worked examples

a 50 cm³ of solution was made using 5.6 g of potassium hydroxide, KOH. What is its concentration in g/dm³ and mol/dm³? *(handwritten: (M/Σ) × 1000 = concentration)*

| 1 cm³ of solution contains (5.6/50) g | so 1 dm³ of solution contains (5.6/50) × 1000 g = 112 g | concentration of solution = **112 g/dm³** |
| 1 mole KOH = (39 + 16 + 1) g = 56 g | 112 g /56 g = 2 mole | concentration of solution = **2 mol/dm³** |

b What is the mass of sodium hydroxide in 100 cm³ of a solution with a concentration of 0.2 mol/dm³?

100 cm³ contains   1 mol NaOH = 40 g   0.02 × 40 = **8 g**
100 × 0.2/1000 mol = 0.02 mol

(handwritten: 1.2/100 × 1000)

▶ 1 **100 cm³ of solution was made using 1.2 g LiOH. What is its concentration in g/dm³ and mol/dm³? (Relative atomic masses: Li = 7, O = 16, H = 1)**
(handwritten: 0.0012 × 1000   12g/dm³)

- Titrations are used to find the volumes of solutions that react exactly.

- If the concentration of one of the solutions is known, and the volumes that react together are known, the concentration of the other solution can be calculated. This information can be used to find the amount of a substance in a sample.

- The concentrations are calculated using balanced symbol equations and moles.

Worked example

A student found that 25.0 cm³ of sodium hydroxide solution with an unknown concentration reacted with exactly 20.0 cm³ of 0.50 mol/dm³ hydrochloric acid. What was the concentration of the sodium hydroxide solution?

The equation for this reaction is: NaOH(aq) + HCl(aq) → NaCl(aq) + H_2O(l)

The concentration of the HCl is 0.50 mol/dm³, so 0.50 mol of HCl are dissolved in 1000 cm³ of acid.

Therefore 20.0 cm³ of acid contains 20 × 0.50/1000 mol = 0.010 mol HCl

The equation for the reaction tells us that 0.010 mol of HCl will react with exactly 0.010 mol of NaOH.

This means that there must have been 0.010 moles of NaOH in the 25.0 cm³ of solution in the conical flask.

So, the concentration of NaOH solution = (0.010/25) × 1000 = **0.40 mol/dm³**

▶ 2 **15.0 cm³ of hydrochloric acid reacted exactly with 25.0 cm³ of sodium hydroxide solution that had a concentration of 0.10 mol/dm³. What was the concentration of the hydrochloric acid in mol/dm³?**

Bump up your grade

To get the highest grades in your exam, you should be able to balance symbol equations for reactions, calculate amounts of substances from titration results and apply these skills to solving problems.

Study tip

Remember this equation for calculations:

$$\text{number of moles} = \frac{\text{mass in grams}}{\text{relative formula mass}}$$

*(handwritten:
7 + 16 + 1 =
24   12/24
= 0.5 mole
0.5 mol /dm³
0.1 mol : 1000
15 × 25 /1000 = 0.0017)*

4.5 Chemical analysis

Key points

- Scientists working in environmental monitoring, medicine and forensic science all need to analyse substances.

- The results of their analysis are often matched against existing databases to identify substances (or suspects in the case of forensics).

Gel electrophoresis plate

Sample added   Position of bands depends on composition of DNA in sample

Analysing a DNA sample

Genetic analysis

- Chemists and other scientists use a variety of methods to analyse substances for many purposes including environmental, medical and forensic investigations.

- They may use traditional 'wet chemistry' methods similar to those you have used in this chapter or instrumental methods such as gas chromatography and mass spectrometry (covered in your Additional Science Revision Guide in *C2 3.8*).

- Some methods, called **qualitative**, are used to find out simply if a substance is in a sample, such as the tests for ions in *C3 4.1* and *C3 4.2*. Other techniques can tell us how much of a substance is in a sample, such as titrations in *C3 4.3* and gas chromatography – mass spectrometry (GC–MS) covered in your Additional Science Revision Guide in *C2 3.8*, and these are known as **quantitative** methods.

> **1** *When is it necessary to use quantitative analysis?*
> to find the quantity of the compounds

- Large databases of the results of analyses have been built up with the aid of computers. These are used to identify substances in samples, to identify individuals, or to monitor changes in amounts of substances over time.

> **2** *Why it is necessary to have a large database for DNA analysis to be used to identify individuals?* Many people are
> There is a larger chance of identification of individual

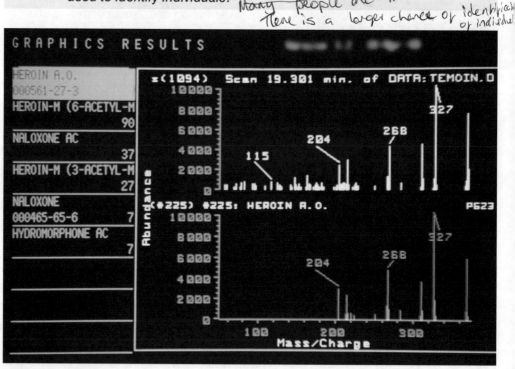

Forensic drug analysis

Study tip

In the exam you may be given results from analyses that have been done in contexts that you have not studied. Do not be put off by different contexts. Try to apply what you have learnt about the chemical tests and titrations you have done.

4.6 Chemical equilibrium

Higher

Key points

- In a reversible reaction the products of the reaction can react to re-form the original reactants. **[H]**

- In a closed system, equilibrium is achieved when the rates of the forward and reverse reactions are equal. **[H]**

- Changing the reaction conditions can change the amounts of products and reactants in a reaction mixture at equilibrium. **[H]**

Study tip

It is the rates of the forward and reverse reactions that are equal at equilibrium, not the amounts of reactants and products. However, the amounts of reactants and products remain constant when the reaction is at equilibrium.

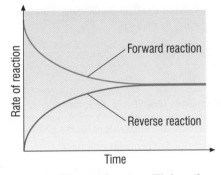

In a reversible reaction at equilibrium, the rate of the forward reaction is equal to the rate of the reverse reaction

Reversible reactions and equilibrium

Some chemical reactions are reversible. This means that the products can react together to make the reactants again:

$$A + B \rightleftharpoons C + D$$

In a closed system no reactants or products can escape. For a reversible reaction in a closed system, **equilibrium** is reached when the rate of the forward reaction is equal to the rate of the reverse reaction. At equilibrium both reactions continue to happen, but the amounts of reactants and products remain constant.

▷ 1 *Explain what is meant by equilibrium.*

balance, a closed system where both sides of the equation are balanced.

Changing the concentration of a reactant or product

The amounts of the reactants and products for a reversible reaction can be changed by changing the reaction conditions. This is important for the chemical industry in controlling reactions. For example, increasing the concentration of a reactant will cause more products to be formed as the system tries to achieve equilibrium. If a product is removed, more reactants will react to try to achieve equilibrium and so more product is formed.

For example, for the reaction: $ICl + Cl_2 \rightleftharpoons ICl_3$

If chlorine is added, the concentration of chlorine is increased and more ICl_3 is produced.

If chlorine is removed, the concentration of chlorine is decreased and more ICl is produced.

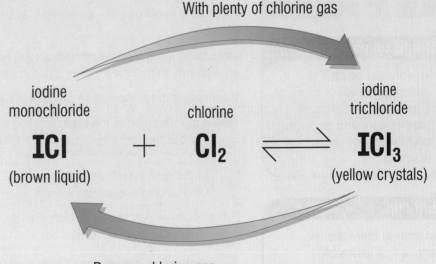

With plenty of chlorine gas

iodine monochloride | chlorine | iodine trichloride

ICl (brown liquid) + **Cl₂** ⇌ **ICl₃** (yellow crystals)

Remove chlorine gas

Changing the equilibrium mixture

▷ 2 *To make SO₃, the reaction $2SO_2(g) + O_2(g) \rightleftharpoons 2SO_3(g)$ is done in a reactor over a heated catalyst. Why is the SO₃ removed from the reactor as soon as it is made?*

More reactants will react to try and achieve equilibrium.

Key word: equilibrium

53

4.7 Altering conditions

- Changing the pressure can affect reversible reactions involving gases at equilibrium.
  - Increasing the pressure favours the reaction with the smaller number of molecules of gas formed.
  - Decreasing the pressure favours the reaction with the larger number of molecules of gas formed.
- Changing the temperature at which we carry out a reversible reaction can change the amount of products formed at equilibrium.
  - Increasing the temperature favours the endothermic reaction.
  - Decreasing the temperature favours the exothermic reaction.

Changing pressure

If we change the conditions of a system at equilibrium, the position of equilibrium shifts as if to try to cancel out the change.

For reversible reactions that have different numbers of molecules of gases on one side of the equation than the other, changing the pressure will affect the position of equilibrium. For example, if the pressure is increased, the position of equilibrium will shift to try to reduce the pressure (favouring the reaction that produces fewer molecules of gas).

This is summarised in the table:

If the forward reaction produces **more** molecules of gas …	If the forward reaction produces **fewer** molecules of gas …
… an increase in pressure decreases the amount of products formed.	… an increase in pressure increases the amount of products formed.
… a decrease in pressure increases the amount of products formed.	… a decrease in pressure decreases the amount of products formed.

For example: in the reversible reaction: $2NO_2(g) \rightleftharpoons N_2O_4(g)$ there are more gaseous reactant molecules than gaseous product molecules. Therefore increasing the pressure will increase the amount of N_2O_4 (product) in the mixture at equilibrium.

> **1** *For the reaction $2SO_2(g) + O_2(g) \rightleftharpoons 2SO_3(g)$, what change in pressure will increase the amount of SO_3 in the equilibrium mixture?*
>
> Increased pressure.

Changing temperature

Reversible reactions are exothermic in one direction and endothermic in the other direction.

Increasing the temperature favours the reaction in the endothermic reaction. The equilibrium shifts as if to lower the temperature by taking in energy.

Decreasing the temperature favours the exothermic reaction.

This is summarised in the table:

If the forward reaction is exothermic …	If the forward reaction is endothermic …
… an increase in temperature decreases the amount of products formed.	… an increase in temperature increases the amount of products formed.
… a decrease in temperature increases the amount of products formed.	… a decrease in temperature decreases the amount of products formed.

For example: for the reversible reaction: $2NO_2(g) \rightleftharpoons N_2O_4(g)$ the forward reaction is exothermic, so increasing the temperature will produce more NO_2 (reactant) in the mixture at equilibrium.

> **2** *The reaction $2SO_2(g) + O_2(g) \rightleftharpoons 2SO_3(g)$ is exothermic in the forward direction. What change in temperature will increase the amount of SO_3 at equilibrium?*
>
> Increase

Student Book
pages 244–245 **C3**

4.8 Making ammonia – the Haber process

- The Haber process is used to manufacture ammonia, which can be used to make fertilisers and other chemicals.
- Nitrogen from the air and hydrogen, which is usually obtained from natural gas, are purified and mixed in the correct proportions.
- The gases are passed over an iron catalyst at a temperature of about 450°C and a pressure of about 200 atmospheres.
- These conditions are chosen to give a fast rate of reaction and a reasonable yield of ammonia.
- The reaction is reversible: $N_2(g) + 3H_2(g) \rightleftharpoons 2NH_3(g)$.

▶ **1** *Write a word equation for the manufacture of ammonia.*
  Nitrogen + Hydrogen ⇌ Ammonia.

- Some of the ammonia that is produced breaks down into nitrogen and hydrogen and the yield of ammonia is only about 15%.
- The gases that come out of the reactor are cooled so the ammonia condenses. The liquid ammonia is separated from the unreacted gases. The unreacted gases are recycled so they are not wasted.

▶ **2** *What is done in the Haber process to conserve raw materials?*
  The unreacted gases are reused.

Key points

- Ammonia is an important chemical for making other chemicals, including fertilisers.
- Ammonia is made from nitrogen and hydrogen in the Haber process.
- The Haber process is done using conditions which are chosen to give a reasonable yield of ammonia as quickly as possible.
- Any unreacted nitrogen and hydrogen are recycled in the Haber process.

Student Book
pages 246–247 **C3**

4.9 The economics of the Haber process

Why there is an optimum pressure for the Haber process
In the Haber process nitrogen and hydrogen react to make ammonia in a reversible reaction:
$$N_2(g) + 3H_2(g) \rightleftharpoons 2NH_3(g)$$
The products have fewer molecules of gas than the reactants, so the higher the pressure the greater the yield of ammonia. However, the higher the pressure the more energy is needed to compress the gas. Higher pressures also need stronger reaction vessels and pipes which increases costs.

A pressure of about 200 atmospheres is often used as a compromise between the costs and the yield. More energy is needed.

▶ **1** *Why do higher pressures increase the costs of an industrial process?*

Why there is an optimum temperature for the Haber process
The forward reaction is exothermic and so the lower the temperature the greater the yield of ammonia. However, the reaction rate decreases as the temperature is lowered and the iron catalyst becomes ineffective so it would take a longer time to produce any ammonia.

Therefore, a compromise temperature of about 450°C is usually used to give a reasonable yield in a short time. It is too slow.

▶ **2** *At a temperature of 100°C and 200 atmospheres pressure the yield of ammonia is 98%. Why is the Haber process not done at this temperature?*

Key points

- The Haber process uses a pressure of around 200 atmospheres to increase the amount of ammonia produced.
- Although higher pressures would produce more ammonia, they would make the chemical plant too expensive to build and run.
- A temperature of about 450°C is used for the reaction. Although lower temperatures would increase the yield of ammonia, it would be produced too slowly.

Study tip
Make sure you know the factors that affect the rates of reactions and understand how they apply to the Haber process.

Higher

handwritten: Fe^{2+} SO_4^{2-}

1 When sodium hydroxide solution was added to a solution a green precipitate formed. When hydrochloric acid and barium chloride solution were added to another sample of the solution a white precipitate formed. Which ions were in the solution?

2 A compound gave a lilac colour in a flame test. Nitric acid and silver nitrate solution were added to a solution of the compound and a yellow precipitate was formed. Name the compound.

3 Dilute hydrochloric acid was added to a green compound. The mixture gave off a gas that turned limewater cloudy and a blue solution was formed. When sodium hydroxide solution was added to the blue solution a blue precipitate was produced. Name the green compound.

handwritten: Copper sulfate   Copper $CuSO_4^{2-}$

4 Explain how a pipette and a burette are used to do a titration.

5 Ammonia is made by the Haber process. The equation for the reaction is:

$$N_2(g) + 3H_2(g) \rightleftharpoons 2NH_3(g)$$

  a What are the raw materials used for the process? *handwritten:* Nitrogen & Hydrogen

  b What conditions are used in the Haber process? *handwritten:* 450°C   200 atmospheres of pressure

  c How is ammonia separated from the unreacted nitrogen and hydrogen? *handwritten:* the gases are cooled and liquid flows

6 12.5 cm³ of 0.10 mol/dm³ hydrochloric acid reacted exactly with 25.0 cm³ of potassium hydroxide solution. What was the concentration in mol/dm³ of the potassium hydroxide solution? *handwritten:* 12.5 & .25   factor 2 0.10 = 0.05 mol/dm³ [H]

7 The reaction $CaCO_3(s) \rightleftharpoons CaO(s) + CO_2(g)$ reaches equilibrium in a closed system. The forward reaction is endothermic. How could the amount of calcium oxide produced by the reaction be increased?   [H]

handwritten: Increase the temperature

Chapter checklist ✓✓✓

Tick when you have:

reviewed it after your lesson	✓	☐ ☐
revised once – some questions right	✓	✓ ☐
revised twice – all questions right	✓	✓ ✓

Move on to another topic when you have all three ticks

Tests for positive ions	☐ ☐ ☐
Tests for negative ions	☐ ☐ ☐
Titrations	☐ ☐ ☐
Titration calculations	☐ ☐ ☐
Chemical analysis	☐ ☐ ☐
Chemical equilibrium	☐ ☐ ☐
Altering conditions	☐ ☐ ☐
Making ammonia – the Haber process	☐ ☐ ☐
The economics of the Haber process	☐ ☐ ☐

Organic chemistry

Key points

- The homologous series of alcohols contain the –OH functional group.
- The homologous series of carboxylic acids contain the –COOH functional group.
- The homologous series of esters contains the –COO– functional group.

Study tip

You do not need to be able to name esters other than ethyl ethanoate, but you should be able to recognise an ester from its name, its structural formula or its displayed formula.

Bump up your grade

You should be able to name the first three alcohols and carboxylic acids and write displayed and structural formulae for them.

$$H-C \overset{=O}{\underset{O-H}{}}$$

5.1 Structures of alcohols, carboxylic acids and esters

- Organic molecules form the basis of living things and all contain carbon atoms. Carbon atoms bond covalently to each other to form the 'backbone' of many series of organic molecules.
- Series of molecules that have a general formula are called **homologous series**. The alkanes and the alkenes are two homologous series made of only hydrogen and carbon atoms.

▶ **1 Name the first three members of the alkanes.**

- **Alcohols** contain the **functional group** –O–H. If one hydrogen atom from each alkane molecule is replaced with an –O–H group, we get a homologous series of alcohols.
- The first three members of this series are methanol, ethanol and propanol.

Methanol   Ethanol   Propanol

The displayed formulae of the first three members of the alcohol series

- A **structural formula** shows which atoms are bonded to each carbon atom and the functional group. The structural formula of ethanol is CH_3CH_2OH.

▶ **2 Write the structural formula of propanol.**

$$CH_3{}^2CH_2 \; OH$$

- **Carboxylic acids** have the functional group –COOH.
- The first three members of the carboxylic acids are methanoic acid, ethanoic acid and propanoic acid. Their structural formulae are $HCOOH$, CH_3COOH and CH_3CH_2COOH.

Methanoic acid   Ethanoic acid   Propanoic acid

The displayed formula of the first three carboxylic acids

▶ **3 Draw the displayed formula of methanoic acid.**

- **Esters** have the functional group –COO–. If the H atom in the –COOH group of a carboxylic acid is replaced by a hydrocarbon group the compound is an ester.
- Ethyl ethanoate has the structural formula $CH_3COOCH_2CH_3$.

$$CH_3COOCH_2CH_3$$

Ethyl ethanoate

The displayed formula of ethyl ethanoate

Key words: homologous series, functional group

5.2 Properties and uses of alcohols

- Alcohols with smaller molecules, such as methanol, ethanol and propanol, mix well with water and produce neutral solutions.
- Many organic substances dissolve in alcohols and so this makes them useful solvents.
- Ethanol is the main alcohol in wine, beer and other alcoholic drinks.

Many organic substances dissolve in them

>1 **Why do many perfumes contain ethanol?** *So it is easy to add fragrance*

- Alcohols burn in air. When burned completely they produce carbon dioxide and water. They are used as fuels, for example in spirit burners or in combustion engines and they can be mixed with petrol.

$$\text{ethanol} + \text{oxygen} \rightarrow \text{carbon dioxide} + \text{water}$$
$$C_2H_5OH + 3O_2 \rightarrow 2CO_2 + 3H_2O$$

- Sodium reacts with alcohols to produce hydrogen gas, but the reactions are less vigorous than when sodium reacts with water.
- Alcohols can be oxidised by chemical oxidising agents such as potassium dichromate to produce carboxylic acids. Some microbes in the air can also oxidise solutions of ethanol to produce ethanoic acid, which turns alcoholic drinks sour and is the main acid in vinegar.

>2 **Ethanol and water are both colourless liquids. Suggest one chemical test you could do to tell them apart.** *lime water test when heat is added.*

Key points

- Alcohols are used as solvents and fuels, and ethanol is the main alcohol in alcoholic drinks.
- Alcohols burn in air, forming carbon dioxide and water.
- Alcohols react with sodium to form a solution and give off hydrogen gas.
- Ethanol can be oxidised to ethanoic acid, either by chemical oxidising agents or by the action of microbes. Ethanoic acid is the main acid in vinegar.

Study tip

You should be able to identify the organic products of reactions from given reagents, but you do not need to be able to write balanced equations for any reactions of organic compounds, except for the combustion of alcohols.

Bump up your grade

You should know the main reactions of alcohols and how to tell if a liquid is an alcohol, an alkane, an acid or water.

Alcohols are used as solvents in perfumes

5.3 Carboxylic acids and esters

Key points

- Solutions of carboxylic acids have a pH value less than 7. Their acidic solutions react with carbonates, gently fizzing as they release carbon dioxide gas.

- Aqueous solutions of weak acids have a higher pH value than solutions of strong acids with the same concentration.   **[H]**

- Esters are made by reacting a carboxylic acid and an alcohol together with an acid catalyst.

- Esters are volatile compounds used in flavourings and perfumes.

Ethanoic acid reacts with carbonates to produce carbon dioxide

- Carboxylic acids dissolve in water to produce solutions with a pH value of less than 7. They have the properties typical of all acids. For example, when carboxylic acids are added to carbonates they fizz because they react to produce carbon dioxide. A salt and water are also produced.

  Their pH is less then 7

> 1 *Why do carboxylic acids have properties similar to all other acids?*

- Carboxylic acids are different from other acids because they react with alcohols in the presence of an acid catalyst to produce esters. For example, ethanol and ethanoic acid react together when mixed with sulfuric acid as a catalyst, to produce ethyl ethanoate and water.

- Esters are volatile compounds and have distinctive smells. Some esters have pleasant fruity smells and are used as flavourings and in perfumes.

> 2 *Why are some esters used as flavourings?* *They have pleasant smells*

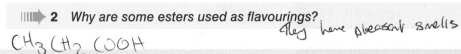

CH_3CH_2COOH

Why carboxylic acids are weak acids

- In aqueous solution, hydrochloric acid ionises completely to hydrogen ions and chloride ions.

$$HCl(aq) \rightarrow H^+(aq) + Cl^-(aq)$$

- Acids that ionise completely in aqueous solutions are known as **strong acids**.

- When ethanoic acid dissolves in water, it does not ionise completely and some of the ethanoic acid molecules remain as molecules in the solution:

$$CH_3COOH(aq) \rightleftharpoons CH_3COO^-(aq) + H^+(aq)$$

- Acids that do not ionise completely in aqueous solution are known as **weak acids**.

- In aqueous solutions of equal concentration, weak acids have a higher pH and react more slowly than strong acids.

> 3 *Write a balanced equation to show that propanoic acid is a weak acid.*

$$CH_3CH_2COOH \rightleftharpoons CH_3CH_2COO + H$$

> **Study tip**
>
> You can tell that a solution is acidic if its pH is less than 7 or if it fizzes when added to a carbonate.

> **Bump up your grade**
>
> If you are taking the Higher Tier paper, you should be able to explain why carboxylic acids are weak acids and know how to tell the difference between weak and strong acids.

Key words: strong acid, weak acid

Higher

Student Book
pages 256–257

C3

5.4 Organic issues

Key points

- Alcohols, carboxylic acids and esters have many uses which benefit society.

- However, some of these substances, such as ethanol and solvents, can be abused.

- In future, the use of biofuels, such as ethanol and esters, could help society as crude oil supplies run out.

- However, future uses of biofuels might conflict with the need to feed the world.

Addiction to alcohol can cause many problems for the individuals themselves and society as a whole

- Alcohols, carboxylic acids and esters are important organic chemicals that can be used in many ways in foods, in drinks, as solvents and as fuels for the benefit of society.

- However, depending on how they are used, there can also be disadvantages. For example, alcoholic drinks and solvents can be abused and this can lead to health and social problems.

- Biofuels offer an alternative to fossil fuels and may help with the problems of diminishing resources and climate change.

- However, some of the crops used for the production of biofuels require the use of agricultural land that could be used to grow food.

- Advantages and disadvantages for any use of resources may change over time or when there are new developments and so monitoring and careful research are needed.

▶ **1** *Why is it important to review our use of organic chemicals?*

Study tip

You should have discussed some of the social and economic benefits and drawbacks of using alcohols, carboxylic acids and esters. You may be given further specific data and information in the exams.

(Precipitation)   + (Flame tests)

Li = crimson
Na = yellow
Potassium = lilac
Ba = green

Ca = red
Copper = green

Positive Sodium hydroxide =   W   Sulfate ions
                                          =
Copper   blue   hcl + bacl
Fe^{2+} = green   white precipitate
Fe^{3+} = brown

Negative   Carbonate ions =

            + hcl

        Halogens
     nitric + silver nitrate =
cl =   white
br =   cream
I =   yellow

1. **Name and give the structural formula of the first three members of the series of alcohols.**

   Methanol ethanol propanol
   CH_3OH CH_3CH_2OH $CH_3CH_2CH_2OH$

2. **Name and give the structural formula of the carboxylic acid with three carbon atoms in its molecule.**

   propanoic acid $CH_3 CH_2 COOH$

3. **Draw the displayed formula of ethyl ethanoate.**

   CH_3 $COO CH_2 CH_3$

4. **What properties make ethanol a useful solvent?**

   It can bond to many organic substances

5. **Describe what happens when a small piece of sodium is added to some ethanol in a beaker.**

   Hydrogen gas will be produced slightly, as well as some sodium hydroxide

6. **A glass of beer containing 5% ethanol was left exposed to the air for 12 hours. The beer turned sour. Explain why.**

   The ethanol was oxidised by some microbes in the air, turning it sour

7. **Describe one reaction of ethanoic acid which is similar to the reactions of all other acids.**

   They fizz when added to carbonates

8. **Ethanol and ethanoic acid can react together to produce an ester. Name the ester and describe the conditions used for the reaction.**

   ethyl ethanoate acid catalyst heat

9. **Suggest why ethyl butanoate is added to some fruit drinks.**

   it has a pleasant smell.

10. **Suggest one advantage and one disadvantage of using ethanol as an alternative fuel to petrol.**

    It takes up farmland, closer to carbon neutral.

11. **You have been given aqueous solutions of hydrochloric acid and ethanoic acid that have the same concentration. Suggest one simple test that you could do to decide which solution is ethanoic acid.** add an open flame and see [H]

    $2CH_3CH_2CH_2OH + 9O_2 \rightarrow 6CO_2$ which combusts $+ 8H_2O$

12. **Write a balanced symbol equation for the complete combustion of propanol.** [H]

    $2 CH_3 CH_2 CH_2 OH \xrightarrow{heat} CO_2 + H$

Chapter checklist		✓ ✓ ✓
Tick when you have:		Structures of alcohols, carboxylic acids and esters ☐ ☐ ☐
reviewed it after your lesson	✓ ☐ ☐	
revised once – some questions right	✓ ✓ ☐	
revised twice – all questions right	✓ ✓ ✓	Properties and uses of alcohols ☐ ☐ ☐
Move on to another topic when you have all three ticks		Carboxylic acids and esters ☐ ☐ ☐
		Organic issues ☐ ☐ ☐

1 The water supplied through the mains in the UK is treated so that it is safe to drink.

   a Which **two** of the following methods are used to treat mains water in the UK?
   chlorination   distillation   evaporation   filtration   *(2 marks)*

   b Which **two** of the following ions cause hardness in water?
   calcium   chloride   magnesium   potassium   sodium   sulfate   *(2 marks)*

   c Describe how you could test some water to find out if it is hard. Give the result of the test.   *(2 marks)*

   d When it is heated, one type of hard water produces scale.
      i What is this type of hard water called?   *(1 mark)*
      ii Why is scale a problem?   *(1 mark)*

   e Give **two** methods that can be used to soften any type of hard water without heating.   *(2 marks)*

   f Suggest **two** reasons why mains water is not softened before it reaches the consumer.   *(2 marks)*

   g Explain, as fully as you can, how scale is formed. You should include at least **one** balanced equation in your answer.   **[H]** *(4 marks)*

Study tip

When asked for a chemical test, as in Q1(c) and Q4, you should always describe how to do the test and give the result of the test.

Study tip

When you are asked for a specific number of answers, for example **two** methods in Q1(e) or **two** reasons in Q1(f) do not give more answers than the number required. If you give additional incorrect answers you will lose marks.

2 Methanol, ethanol and propanol are the first three members of a series of compounds.

   a What general name is used for members of this series?   *(1 mark)*

   b Write the structural formula of ethanol.   *(1 mark)*

   c What is the functional group that all members of this series have?   *(1 mark)*

   d *In this question you will be assessed on using good English, organising information clearly and using specialist terms where appropriate.*

   A student wanted to compare the amount of energy released when methanol, ethanol and propanol were burned. The student used the apparatus shown in the diagram.

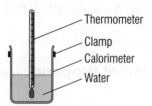

   The student used the equation $Q = mc\Delta T$ to calculate the energy released by each compound, with $c = 4.2\ \text{J/g}\,^{\circ}\text{C}$.

   Describe how the student should use this apparatus and the measurements that he should make to compare the energy released by the three compounds.   *(6 marks)*

   e Sketch an energy level diagram for the combustion of ethanol showing the activation energy and the energy change of the reaction.   *(4 marks)*

3 The diagram shows the Haber process for making ammonia.

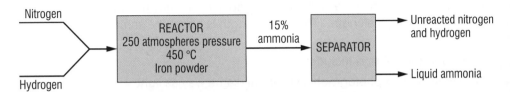

The equation for the reaction is: $N_2(g) + 3H_2(g) \rightleftharpoons 2NH_3(g)$

a What is the source of the nitrogen used in the process? *(1 mark)*

b Give **one** source of the hydrogen used in the process. *(1 mark)*

c Why is iron powder used in the reactor? *(1 mark)*

d What happens to the unreacted nitrogen and hydrogen? *(1 mark)*

e How are the conditions changed in the separator so the ammonia liquefies? *(1 mark)*

f What does the symbol $\rightleftharpoons$ tell you about the reaction? *(1 mark)*

g Explain why a high pressure is used for the reaction. **[H]** *(2 marks)*

h The yield of ammonia decreases with an increase in temperature. Explain why.
[H] *(2 marks)*

i A relatively high temperature of 450 °C is used in the process. Explain why.
[H] *(2 marks)*

4 For each of the following pairs of substances suggest **one** test that you could do to tell them apart. Give the results of the test for both substances.

a sodium carbonate and sodium nitrate *(2 marks)*

b potassium chloride and potassium iodide *(2 marks)*

c calcium chloride and magnesium chloride *(2 marks)*

d iron(II) sulfate and iron(III) sulfate *(2 marks)*

e ethanol and ethanoic acid *(2 marks)*

5 The table shows some bond energies. **[H]**

Bond	Bond energy in kJ per mole
C–H	413
C–C	347
O=O	498
C=O	805
H–O	464

Pentane burns in air. The equation for the reaction is: $C_5H_{12} + 8O_2 \rightarrow 5CO_2 + 6H_2O$

a Use the bond energies in the table to calculate the energy change for this reaction. Show all your working. *(5 marks)*

b Explain, in terms of bond energies, why this reaction is exothermic. *(1 mark)*

Student Book
pages 208–209 **P3**

1.1 X-rays

Student Book
pages 210–211 **P3**

Student Book
pages 208–209

Key points

- X-rays are used in hospitals:
  - to make images and CT scans
  - to destroy tumours at or near the body surface.
- X-rays can damage living tissue when they pass through it.
- X-rays are absorbed more by bones and teeth than by soft tissue.

Bump up your grade

Remember that X-rays can cause cancer but they can also be used to treat cancer.

Key words: X-ray, charge-coupled device (CCD), CT scanner

- **X-rays** are part of the electromagnetic spectrum. They have a high frequency and a very short wavelength. (Their wavelength is about the same size as the diameter of an atom.)
- Properties of X-rays include:
  - they affect a photographic film in the same way as light
  - they are absorbed by metal and bone
  - they are transmitted by healthy tissue.
- X-rays are used to form images of bones on photographic film to check for fractures and dental problems.
- **Charge-coupled devices (CCDs)** can be used to form electronic images of X-rays. **CT scanners** use X-rays to produce digital images of a cross-section through the body. Some body organs made of soft tissue, such as the intestines, can be filled with a contrast medium that absorbs X-rays so that they can be seen on an X-ray image.
- X-rays cause ionisation and can damage living tissue when they pass through it, therefore precautions must be taken when using them. Workers should wear film badges and when possible use lead screens to shield them from the X-rays.
- X-rays may also be used for therapy. They can be used to treat cancerous tumours at or near the body surface.

> **1** *Why do workers in X-ray departments wear lead aprons?*

1.2 Ultrasound

Key points

- Ultrasound waves are sound waves of frequency above 20 000 Hz.
- Ultrasound can be used for diagnosis and treatment.
- Ultrasound waves are partly reflected at a boundary between two different types of body tissue.
- An ultrasound scan is non-ionising so it is safer than an X-ray.

Study tip

You may be asked to do calculations using data from oscilloscope traces.

Key word: ultrasound wave

- The human ear can detect sound waves with frequencies between 20 Hz and 20 000 Hz. Sound waves of a higher frequency than this are called **ultrasound waves**.
- Electronic systems can be used to produce ultrasound waves. When a wave meets a boundary between two different materials, part of the wave is reflected. The wave travels back through the material to a detector. The time it takes to reach the detector can be used to calculate how far away the boundary is. The results may be processed by a computer to give an image.

> **1** *What is the minimum frequency of an ultrasound wave?*

- The distance travelled by an ultrasound pulse can be calculated using the equation
$$s = v \times t$$
Where:
s is the distance travelled in metres, m
v is the speed of the ultrasound wave in metres per second, m/s
t is the time taken in seconds, s.
- In the time between a transmitter sending out a pulse of ultrasound and it returning to a detector, it has travelled from the transmitter to a boundary and back, i.e. twice the distance to the boundary.
- Ultrasound can be used in medicine for scanning. It is non-ionising, so is safer to use than X-rays. It can be used for scanning unborn babies and soft tissue such as the eye. Ultrasound may also be used in therapy, for example to shatter kidney stones into small pieces.

Student Book
pages 212–213

P3

1.3 Refractive index

- Refractive index, n, is a measure of how much a substance can refract a light ray.

- $n = \dfrac{\sin i}{\sin r}$

Study tip

Remember that angles i and r are measured between the ray and the normal.

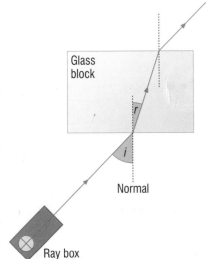

Glass
block

r

i

Normal

Ray box

Refraction of light

- **Refraction** is the change of direction of light as it passes from one transparent substance into another.
- Refraction takes place because waves change speed when they cross a boundary. The change in speed of the waves causes a change in direction, unless the waves are travelling along a normal.

1 What is refraction?

- A light ray will refract when it crosses from air into glass. It is refracted towards the normal.
- The **refractive index** of a substance is a measure of how much the substance can refract a light ray.
- The refractive index is given by the equation:

$$n = \frac{\sin i}{\sin r}$$

Where:
n is the refractive index of the substance
$\sin i$ is the sine of the angle of incidence
$\sin r$ is the sine of the angle of refraction.

2 Which way does a light ray change direction when it crosses from glass to air?

Maths skills

A ray of light travels from air into glass. The angle of incidence is 45° and the angle of refraction is 28°.

Calculate the refractive index of the glass.

$$n = \frac{\sin i}{\sin r}$$

$$n = \frac{\sin 45°}{\sin 28°}$$

$$n = \frac{0.71}{0.47}$$

$$n = 1.5$$

Refractive index is a ratio, so it does not have a unit.

Bump up your grade

A ray of light travelling along a normal is not refracted.

Key words: refraction, refractive index

1.4 The endoscope

Study tip

Remember that total internal reflection only takes place for a ray travelling from a more dense to a less dense material, e.g. from glass into air.

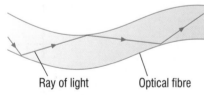

Light rays in an optical fibre

- A light ray will refract when it crosses from glass to air. It is refracted away from the normal. A partially reflected ray is also seen. If the angle of incidence in the glass is gradually increased, the angle of refraction increases until the refracted ray emerges along the boundary. This angle of incidence is called the **critical angle**, c.

- If the angle of incidence is increased beyond the critical angle the light ray undergoes **total internal reflection**. When total internal reflection occurs, the angle of reflection is equal to the angle of incidence.

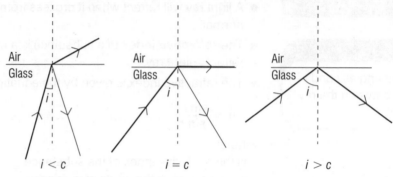

Refraction and total internal reflection

The critical angle is related to the refractive index by the equation:

$$n = \frac{1}{\sin c}$$

Where:
n is the refractive index
c is the critical angle.

> **1** *What is the critical angle for glass of refractive index 1.5?*

- An **endoscope** is a device used to look inside a patient's body without cutting it open or when performing keyhole surgery. The endoscope contains bundles of **optical fibres**. These are very thin, flexible glass fibres. Visible light can be sent along the fibres by total internal reflection.

- Laser light may be used as an energy source in an endoscope to carry out some surgical procedures such as cutting, cauterising and burning. The colour of the laser light is matched to the type of tissue to produce maximum absorption. Eye surgery on the retina in the eye can be carried out by using laser light that passes straight through the cornea at the front of the eye but is absorbed by the retina at the back.

> **2** *What is an optical fibre?*

Key words: critical angle, total internal reflection, endoscope, optical fibre

1.5 Lenses

Key points

- A converging lens focuses parallel rays to a point called the principal focus.

- A diverging lens makes parallel rays spread out as if they came from a point called the principal focus.

- A real image is formed by a converging lens if the object is further away than the principal focus.

- A virtual image is formed by a diverging lens, and by a converging lens if the object is nearer to the lens than the principal focus.

- Magnification = $\dfrac{\text{image height}}{\text{object height}}$

Study tip

Make sure you don't confuse the terms 'converging' and 'diverging'.

Converging (convex) lens

- Parallel rays of light that pass through a **converging (convex) lens** are refracted so that they converge to a point. This point is called the **principal focus** (focal point). The distance from the centre of the lens to the principal focus is the **focal length**.

- Because light can pass through the lens in either direction, there is a principal focus on either side of the lens.

- If the object is further away from the lens than the principal focus, an inverted, **real image** is formed. The size of the image depends on the position of the object. The nearer the object is to the lens, the larger the image.

- If the object is nearer to the lens than the principal focus, an upright, **virtual image** is formed behind the object. The image is magnified – the lens acts as a **magnifying glass**.

- The **magnification** can be calculated using:

$$\text{magnification} = \frac{\text{image height}}{\text{object height}}$$

Diverging (concave) lens

- Parallel rays of light that pass through a **diverging (concave) lens** are refracted so that they diverge away from a point. This point is called the principal focus.

- The distance from the centre of the lens to the principal focus is the focal length.

- Because light can pass through the lens in either direction, there is a principal focus on either side of the lens.

- The image produced by a diverging (concave) lens is always virtual.

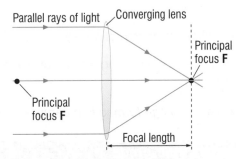

The focal length of a converging lens

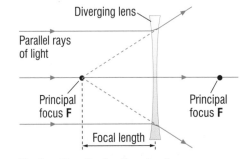

The focal length of a diverging lens

> 1 What is the principal focus of a converging (convex) lens?
> 2 How can a converging lens be made to produce a virtual image?

- The symbols below can be used in ray diagrams to represent lenses.

Key words: converging lens, principal focus, focal length, real image, virtual image, magnifying glass, magnification, diverging lens

1.6 Using lenses

Key points

- A ray diagram can be drawn to find the position and nature of an image formed by a lens.

- When an object is placed between a converging lens and F, the image formed is virtual, upright, magnified and on the same side of the lens as the object.

- A camera contains a converging lens that is used to form a real image of an object.

- A magnifying glass is a converging lens that is used to form a virtual image of an object.

- We can draw ray diagrams to find the image that different lenses produce with objects in different positions.

- The line through the centre of the lens and at right angles to it is called the **principal axis**. Include this in your diagram.

- Ray diagrams use three construction rays from a single point on the object to locate the corresponding point on the image:
  - A ray parallel to the principal axis is refracted through the principal focus.
  - A ray through the centre of the lens travels straight on, without refraction.
  - A ray through the principal focus is refracted parallel to the principal axis.

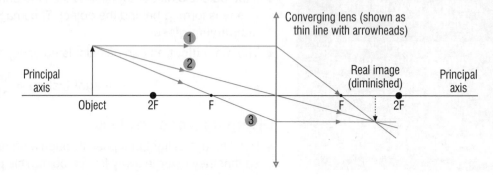

Ray ① is parallel to the axis and is refracted through F

Ray ② passes straight through the centre of the lens

Ray ③ passes through F and is refracted parallel to the axis

Formation of a real image by a converging lens

▷ **1** *What are construction rays?*

- A camera uses a converging lens to form a real image of an object on a film or an array of CCDs.

▷ **2** *Is the image formed in a camera real or virtual?*

⚠ Bump up your grade

Make sure you practise drawing ray diagrams. Only two of the construction rays are needed to find the image, but if you have time it is worth drawing all three to be sure that you have the correct position.

Key word: principal axis

Student Book
pages 220–221

P3

1.7 The eye

Key points

- Light is focused on to the retina by the cornea and the eye lens, which is a variable focus lens.
- The normal human eye has a range of vision from 25 cm to infinity.
- $P = \dfrac{1}{f}$

Study tip

When calculating the power of a lens, make sure that the focal length is in metres so that the power of the lens is in dioptres.

Inside the eye

- Light enters the eye through the **cornea**. The cornea and the **eye lens** focus the light on to the **retina**. The **iris** adjusts the size of the **pupil** to control the amount of light entering the eye.
- The **ciliary muscles** alter the thickness of the lens to control the fine focusing of the eye. They are attached to the lens by the **suspensory ligaments**.

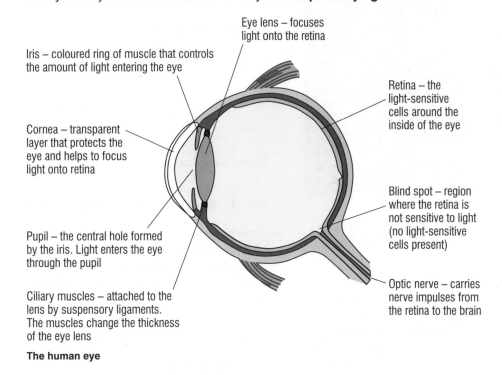

Eye lens – focuses light onto the retina

Iris – coloured ring of muscle that controls the amount of light entering the eye

Cornea – transparent layer that protects the eye and helps to focus light onto retina

Retina – the light-sensitive cells around the inside of the eye

Blind spot – region where the retina is not sensitive to light (no light-sensitive cells present)

Pupil – the central hole formed by the iris. Light enters the eye through the pupil

Ciliary muscles – attached to the lens by suspensory ligaments. The muscles change the thickness of the eye lens

Optic nerve – carries nerve impulses from the retina to the brain

The human eye

1 *Which two structures focus the light entering the eye?*

- The normal human eye has a **near point** of 25 cm and a **far point** of infinity, so its **range of vision** is from 25 cm to infinity.

Lens power

- The **power of a lens** is given by:

$$P = \frac{1}{f}$$

Where:
P is the power of the lens in **dioptres**, D
f is the focal length of the lens in metres, m.

2 *What is the power of a lens of focal length 0.25 m?*

Key words: near point, far point, range of vision, power of a lens, dioptre

1.8 More about the eye

Key points

- A short-sighted eye is an eye that can only see near objects clearly. We use a diverging (concave) lens to correct it.

- A long-sighted eye is an eye that can only see distant objects clearly. We use a converging (convex) lens to correct it.

- The higher the refractive index of the glass used to make a corrective lens, the flatter and thinner the lens can be.   [H]

Bump up your grade

You may be asked to compare the structure of the eye and the camera. The camera has a lens of fixed shape but variable position. The eye has a lens of variable shape but fixed position.

- A person with **short sight** can see close objects clearly, but distant objects are blurred because the uncorrected image is formed in front of the retina. Short sight is caused by the eyeball being too long or the eye lens being too powerful. Short sight may be corrected using a diverging lens.

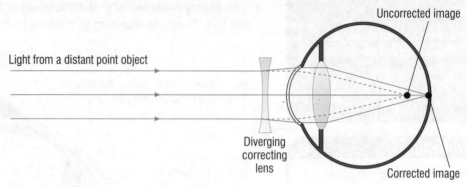

Short sight and its correction

▐▐▐▶ **1** *Which type of lens may be used to correct short sight?*

- A person with **long sight** can see distance objects clearly, but close objects are blurred because the uncorrected image is formed behind the retina. Long sight is caused by the eyeball being too short or the eye lens being too weak. Long sight may be corrected using a converging lens.

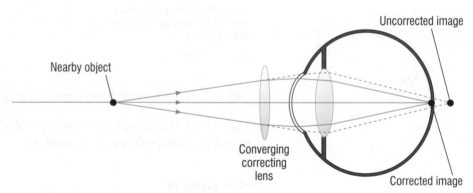

Long sight and its correction

- The focal length of a lens is determined by:
  - the refractive index of the material from which the lens is made
  - the curvature of the two surfaces of the lens.

For a lens of a given focal length, the greater the refractive index of the lens material, the flatter and thinner the lens can be manufactured.

▐▐▐▶ **2** *What defect of the eyeball may cause it to be long-sighted?*

Key words: short sight, long sight

1 Why do workers in hospital X-ray departments wear film badges?

2 What is a CT scanner?

3 What is an ultrasound wave?

4 Why are X-rays not normally used to produce an image of an unborn baby?

5 A ray of light travels through glass of refractive index 1.54. The angle of incidence is 15°. What is the angle of refraction? [H]

6 What happens to a ray of light that enters a glass block along a normal?

7 A ray of light strikes the boundary between glass and air at the critical angle. What will happen to the ray?

8 What is the principal focus of a diverging lens?

9 Describe the image formed by a diverging lens.

10 What are the ciliary muscles?

11 What is the power of a lens of focal length 16 cm?

12 A lens used as a magnifying glass produces a magnification of 6. If the height of the image is 9 cm, what is the height of the object?

Chapter checklist ✓✓✓

Tick when you have:

reviewed it after your lesson ✓ ☐ ☐

revised once – some questions right ✓ ✓ ☐

revised twice – all questions right ✓ ✓ ✓

Move on to another topic when you have all three ticks

	✓	✓	✓
X-rays	☐	☐	☐
Ultrasound	☐	☐	☐
Refractive index	☐	☐	☐
The endoscope	☐	☐	☐
Lenses	☐	☐	☐
Using lenses	☐	☐	☐
The eye	☐	☐	☐
More about the eye	☐	☐	☐

Student Book
pages 226–227

P3

2.1 Moments

Key points

- The moment of a force is a measure of the turning effect of the force on an object.
- $M = F \times d$
- To increase the moment of a force F, increase F or increase d.

A turning effect

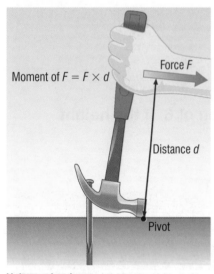

Moment of $F = F \times d$

Force F

Distance d

Pivot

Using a claw hammer

- The turning effect of a force is called its **moment**.
- The size of the moment is given by the equation:

$$M = F \times d$$

Where:

M is the moment of the force in newton-metres, Nm

F is the force in newtons, N

d is the perpendicular distance from the **line of action** of the force to the **pivot** in metres, m.

> **1** A door opens when you apply a force of 20 N at right angles to it, 0.6 m from the hinge. What is the moment of the force about the hinge?
>
> **2** What force would be needed to open the door if it were applied 0.3 m from the hinge? **[H]**

- To increase the moment:
  - either the force must increase
  - or the distance to the pivot must increase.
- It is easier to undo a wheel-nut by pushing on the end of a long spanner than a short one. That's because the long spanner increases the distance between the line of action of the force and the pivot.
- We make use of a lever to make a job easier. When using a lever, the force we are trying to move is called the **load** and the force applied to the lever is the **effort**. A lever acts as a force multiplier, so the effort we apply can be much less than the load.

Study tip

In examination questions, moments may be applied to lots of different situations such as:
- opening a door or a can of paint
- moving something heavy in a wheelbarrow, or
- using a crowbar or a spanner.

The idea is always the same; for any particular force, make the distance to the pivot bigger to make the moment bigger.

Bump up your grade

Notice that the moment equation uses the term 'perpendicular distance'. This means the shortest distance from the line that the force acts along.

Key words: moment, line of action, pivot, load, effort

Student Book
pages 228–229 **P3**

2.2 Centre of mass

- Although any object is made up of many particles, its mass can be thought of as being concentrated at one single point. This point is called the **centre of mass**.
- Any object that is freely suspended will come to rest with its centre of mass directly below the point of suspension. The object is then in **equilibrium**.

||||▶ **1** *What is the centre of mass of an object?*

- You can find the centre of mass of a thin irregular sheet of a material as follows:
  - Suspend the thin sheet from a pin held in a clamp stand. Because it is freely suspended, it is able to turn.
  - When it comes to rest, hang a plumbline from the same pin.
  - Mark the position of the plumbline against the sheet.
  - Hang the sheet with the pin at another point and repeat the procedure.
  - The centre of mass is where the lines that marked the position of the plumbline cross.
- The position of the centre of mass depends on the shape of the object, and sometimes lies outside the object.
- For a symmetrical object, its centre of mass is along the axis of symmetry. If the object has more than one axis of symmetry, the centre of mass is where the axes of symmetry meet.

||||▶ **2** *Where is the centre of mass of a symmetrical object?*

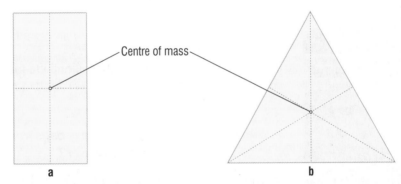

Symmetrical objects

Key points

- The centre of mass of an object is that point where its mass can be thought to be concentrated.
- When a suspended object is in equilibrium, its centre of mass is directly beneath the point of suspension.
- The centre of mass of a symmetrical object is along the axis of symmetry.

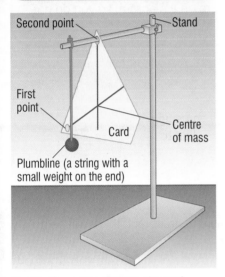

Finding the centre of mass of a card

Study tip

Make sure that you can describe the experiment to find the centre of mass of a thin sheet of a material, including sketching a labelled diagram.

Key words: centre of mass, equilibrium

Student Book
pages 230–231

P3

2.3 Moments in balance

- If an object is in equilibrium it is balanced, not turning. We can take the moments about *any* point and will find that the total clockwise moment and the total anticlockwise moment are equal.
- There are lots of everyday examples of the **principle of moments**, such as seesaws and balance scales.

▶ **1** *If someone sits in the centre of a seesaw, the moment about the pivot is zero. Why?*

▶ **2** *Aimie sits 2m from the centre of a seesaw. Leo weighs twice as much as Aimie. How far from the centre must he sit to balance the seesaw?* **[H]**

Bump up your grade

Be sure to add together all the clockwise moments and all the anticlockwise moments. It may help to tick them off if they are on a diagram, so you do not miss any out. **[H]**

Key word: principle of moments

Student Book
pages 232–233

P3

2.4 Stability

- The line of action of the weight of an object acts through its centre of mass.

If the line of action of the weight lies outside the base of an object, there will be a **resultant moment** and the object will tend to topple over.

▶ **1** *Why does hanging heavy bags from the handle of a pushchair make it more likely to topple over?*

- The wider the base of an object, and the lower its centre of mass, the further it has to tilt before the line of action of the weight moves outside the base. So the stability of an object is increased by making its base wider and its centre of mass lower.

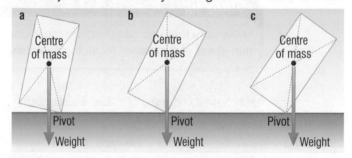

Tilting and toppling a tilted, b at balance, c toppled over

▶ **2** *Why do ten-pin bowling pins have a narrow base and a high centre of gravity?*

Key word: resultant moment

2.5 Hydraulics

Key points

- $P = \dfrac{F}{A}$
- The pressure in a fluid acts equally in all directions.
- A hydraulic system uses the pressure in a fluid to exert a force.

Study tip

Remember that a hydraulic pressure system is usually used as a force multiplier. So if you calculate that the force produced by such a system is less than the effort applied to the system you have made a mistake.

- **Pressure** is given by the equation:

$$P = \frac{F}{A}$$

Where:
P is the pressure in pascals, Pa (or N/m²)
F is the force in newtons, N
A is the cross-sectional area at right angles to the direction of the force in metres squared, m².

 1 *What is the pressure exerted on the ground by a person of weight 300 N if the area of their feet in contact with the ground is 0.04 m²?*

- Liquids are virtually incompressible and the pressure in a liquid is transmitted equally in all directions. This means that a force exerted at one point on a liquid will be transmitted to other points in the liquid. This is made use of in **hydraulic pressure** systems.
- The force exerted by a hydraulic pressure system depends on:
    - the force exerted on the system
    - the area of the cylinder on which this force acts on
    - the area of the cylinder that exerts the force.
- The use of different cross-sectional areas on the effort and load sides of a hydraulic system means that the system can be used as a **force multiplier**. Therefore, a small effort can be used to move a large load.

Maths skills

In a hydraulic pressure system, a force of 25 N is applied to a piston of area 0.50 m². The area of the other piston is 1.5 m². Calculate the pressure transmitted through the system and the force exerted on the other piston.

$P = \dfrac{F}{A}$

$P = \dfrac{25\,\text{N}}{0.5\,\text{m}^2}$

$P = 50\,\text{Pa}$

Pressure transmitted is 50 Pa.

$F = P \times A$

$F = 50\,\text{Pa} \times 1.5\,\text{m}^2$

$F = 75\,\text{N}$

Force exerted on the other piston is 75 N.

2 *What properties of a liquid make it useful in a hydraulic system?*

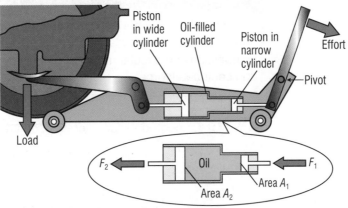

A hydraulic car jack

Key words: pressure, hydraulic pressure

Student Book
pages 236–237

P3

2.6 Circular motion

Key points

- The velocity of an object moving in a circle at constant speed is continually changing as the object's direction is continually changing.
- Centripetal acceleration is the acceleration towards the centre of the circle of an object that is moving round the circle.
- The centripetal force on an object depends on its mass, its speed and the radius of the circle.

- When an object moves in a circle it is continuously changing direction, so it is continuously changing velocity. In other words, it is accelerating. This acceleration is called the **centripetal acceleration**.
- An object only accelerates when a resultant force acts on it. This force is called the **centripetal force** and always acts towards the centre of the circle.
- If the centripetal force stops acting, the object will continue to move in a straight line at a tangent to the circle.
- The centripetal force needed to make an object perform circular motion increases as:
  - the mass of the object increases
  - the speed of the object increases
  - the radius of the circle decreases.

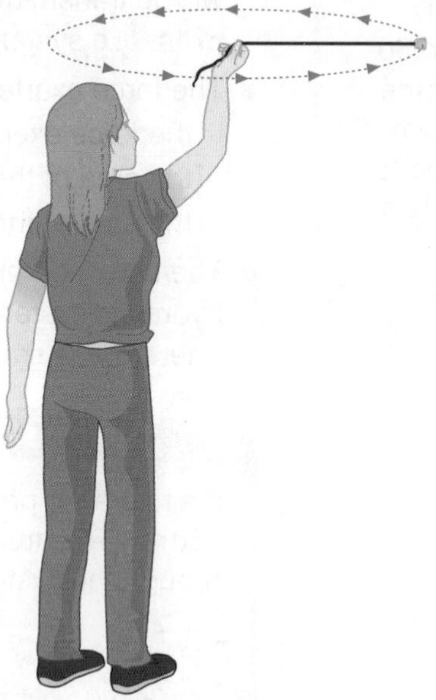

Whirling an object around

1 *A student is whirling a conker around on a piece of string, in a horizontal circle. What force provides the centripetal force?*

2 *What will happen to the conker if the string breaks?*

Study tip

Centripetal force is not a force in its own right. It is always provided by another force, for example gravitational force, electric force or tension.
In questions on circular motion, you may need to identify the force that provides the centripetal force.

Key words: centripetal acceleration, centripetal force

2.7 The pendulum

- A pendulum moves to and fro along the same line. This is an example of **oscillating motion**.
- A **simple pendulum** consists of a mass, called a bob, suspended on the end of a string. When the bob is displaced to one side and let go, the pendulum oscillates back and forth, through the equilibrium position. (The equilibrium position is the position of the pendulum when it stops moving).
- The **amplitude** of the oscillation is the distance from the equilibrium position to the highest position on either side.
- The **time period** of the oscillation is the time taken for one complete cycle, this is:
  - the time taken from the highest position on one side to the highest position on the other side and back to the start position, or
  - the time taken between successive passes in the same direction through the equilibrium position.
    To measure the time period of a pendulum, we can measure the average time for 20 oscillations and divide the timing by 20.
- The time period depends only on the length of the pendulum and increases as its length increases.
- The frequency of the oscillations is the number of complete cycles of oscillation per second.
- The time period and frequency are related by the equation:

$$T = \frac{1}{f}$$

Where:
T is the time period in seconds, s
f is the frequency in hertz, Hz.

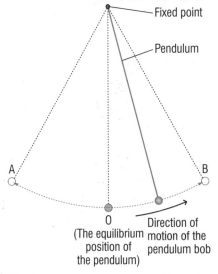

The pendulum

300×10^6 M/S

Maths skills

What is the time period of a pendulum that completes 20 oscillations in 5.0 seconds?

There are $\frac{20.0}{5.0} = 4.0$ oscillations in 1 second, so the frequency is 4.0 Hz.

$$T = \frac{1}{f}$$
$$T = \frac{1}{4.0\,Hz}$$
$$T = 0.25\,s$$

The time period is 0.25 seconds.

- A playground swing is an example of an oscillating motion. If not pushed repeatedly, the swing will come to rest. This is because energy is transferred due to friction at the top of the swing and due to air resistance.

1 What does the time period of a simple pendulum depend on?
2 What is the time period of a pendulum of frequency 10 Hz?

Key words: oscillating motion, simple pendulum, amplitude, time period

1 What is the moment of a force?

2 Why is it easier to move a big rock with a crowbar than with your hands?

3 Where is the position of the centre of mass of each of the shapes below?

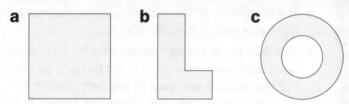

a   b   c

4 Why must an object moving in a circle be accelerating?

5 A force of 15 N is applied at right angles to the end of a spanner 0.5 m long. What is the moment of this force?

6 An object that is freely suspended is displaced slightly then let go. Where will its centre of mass be when it comes to rest?

7 What is the pressure exerted on the ground by a block of weight 20 N if the area in contact with the ground is 0.008 m²?

8 What is a centripetal acceleration?

9 Why will an oscillating simple pendulum eventually come to rest?

10 What is the frequency of a pendulum with a time period of 0.02 s?

11 What is the amplitude of oscillation of a simple pendulum?

12 When will an object topple?

Chapter checklist ✓ ✓ ✓

Tick when you have:

reviewed it after your lesson	✓	☐	☐	Moments	☐	☐	☐
revised once – some questions right	✓	✓	☐	Centre of mass	☐	☐	☐
revised twice – all questions right	✓	✓	✓	Moments in balance	☐	☐	☐
				Stability	☐	☐	☐

Move on to another topic when you have all three ticks

Hydraulics ☐ ☐ ☐

Circular motion ☐ ☐ ☐

The pendulum ☐ ☐ ☐

3.1 Electromagnets

About magnets

- The ends of a magnet are called **magnetic poles**. There is a **north pole** at one end and a **south pole** at the other. The region around the magnet, in which a piece of iron or steel will be attracted to it, is called its **magnetic field**. Iron filings placed near a magnet will form a pattern of lines that loop from one pole to the other. These are **lines of force** or magnetic **field lines**. A plotting compass placed in the magnetic field will always point along a field line.

- If two magnets are brought close to each other with like poles together (north and north or south and south) they will repel each other.

- If unlike poles are together (north and south or south and north) they will attract each other.

> **1** *What is a magnetic field?*

Electromagnets

- When a current flows through a wire, a magnetic field is produced around the wire. An electromagnet is made by wrapping insulated wire around a piece of iron, called the core. When a current flows through the wire the iron becomes strongly magnetised. When the current is switched off the iron loses its magnetism. This temporary magnetism makes electromagnets very useful.

- Electromagnets are used in devices such as scrapyard cranes, circuit breakers, electric bells and relays.

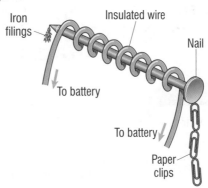

Iron filings · Insulated wire · Nail · To battery · To battery · Paper clips

A simple electromagnet

> **2** *What does an electromagnet consist of?*

Study tip

In the exam you may be given a diagram of an appliance that contains an electromagnet and asked to explain how it works.

Key words: magnetic pole, north pole, south pole, line of force, field line

Key points

- The force between two magnets: like poles repel; unlike poles attract.

- A magnetic field line is the line along which a plotting compass points.

- An electromagnet consists of a coil of insulated wire wrapped round an iron core.

- Electromagnets are used in scrapyard cranes, circuit breakers, electric bells and relays.

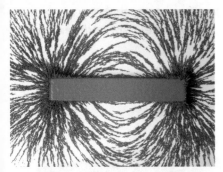

The magnetic field near a bar magnet

3.2 The motor effect

Key points

- In the motor effect, the force:
  - is increased if the current or the strength of the magnetic field is increased,
  - is at right angles to both the direction of the magnetic field and to the wire,
  - is reversed if the direction of either the current or the magnetic field is reversed.
- An electric motor has a coil which turns when a current is passed through it.

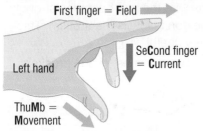

First finger = **F**ield

Left hand

Se**C**ond finger = **C**urrent

Thu**M**b = **M**ovement

Fleming's left-hand rule

Study tip

Make sure you practise using Fleming's left-hand rule.

- When we place a wire carrying an electric current in a magnetic field, it may experience a force. This is called the **motor effect**.
- The force is a maximum if the wire is at an angle of 90° to the magnetic field, and zero if the wire is parallel to the magnetic field.
- Fleming's left-hand rule is used to determine the direction of the force. The thumb and first two fingers of the left hand are all held at right angles to each other:
  - the first finger represents the magnetic field (pointing north to south)
  - the second finger represents the current (pointing positive to negative)
  - the thumb represents the direction of the force.
- The size of the force can be increased by:
  - increasing the strength of the magnetic field
  - increasing the size of the current.
- The direction of the force on the wire is reversed if either the direction of the current or the direction of the magnetic field is reversed.
- The motor effect is used in different appliances.

> **1** *What happens to the direction of the force on a wire carrying a current if the direction of the current and the magnetic field are both reversed?*

The electric motor

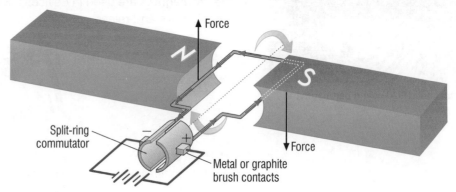

Force

Split-ring commutator

Metal or graphite brush contacts

Force

The electric motor

- The diagram shows a simple electric motor.
  - The speed of the motor is increased by increasing the size of the current.
  - The direction of the motor can be reversed by reversing the direction of the current.
- When a current passes through the coil, the coil spins because:
  - a force acts on each side of the coil due to the motor effect
  - the force on one side of the coil is in the opposite direction to the force on the other side.
- The **'split-ring' commutator** reverses the direction of the current around the coil every half-turn. Because the sides swap over each half-turn, the coil is always pushed in the same direction.

> **2** *The ends of the coil in a motor are parallel to the magnetic field. What is the size of the force on them?*

Key words: motor effect, split-ring commutator

Student Book
pages 246–247

P3

3.3 Electromagnetic induction

- If an electrical conductor 'cuts' through magnetic field lines, a potential difference (pd) is induced across the ends of the conductor.
- If a magnet is moved into a coil of wire, a pd is induced across the ends of the coil. This process is called **electromagnetic induction**. If the wire or coil is part of a complete circuit, a current passes through it.

Key points

- Electromagnetic induction is the process of creating a potential difference using a magnetic field.
- When a conductor cuts the lines of a magnetic field, a potential difference is induced across the ends of the conductor.
- When an electromagnet is used, it needs to be switched on or off to induce a pd.

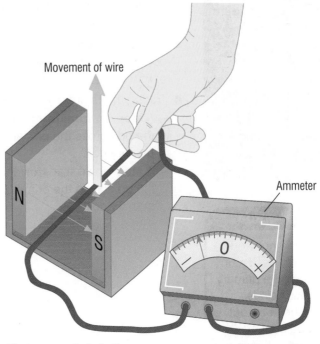

Electromagnetic induction

- If the direction of movement of the wire or coil is reversed, or the polarity of the magnet is reversed, the direction of the induced pd is also reversed. A pd is only induced while there is movement.

> **1** *Why is there no potential difference induced when a bar magnet is held stationary inside a coil of wire?*

- The size of the induced pd is increased by increasing:
  - the speed of movement
  - the strength of the magnetic field
  - the number of turns on the coil.

> **2** *What is the effect on the induced pd of reversing the direction of the current in a conductor cutting magnetic field lines?*

◢ **Bump up your grade**

Remember a potential difference is induced only when the wire or coil and the magnetic field move relative to each other.

Key word: electromagnetic induction

Student Book
pages 248–249

P3

3.4 Transformers

Key points

- A transformer only works on ac because a changing magnetic field is necessary to induce ac in the secondary coil.

- A transformer has an iron core unless it is a switch mode transformer which has a ferrite core.

- A switch mode transformer is lighter and smaller than an ordinary transformer. It operates at high frequency.

Bump up your grade

Transformers do not work with dc, but only ac. If a dc passes through the primary coil a magnetic field is produced in the core, but it would not be continually expanding and collapsing, so no pd would be induced in the secondary coil.

- A **transformer** consists of two coils of insulated wire, called the primary coil and the secondary coil. These coils are wound on to the same iron core. When an alternating current passes through the primary coil, it produces an alternating magnetic field in the core. This field continually expands and collapses.

- The alternating magnetic field lines pass through the secondary coil and induce an alternating potential difference across its ends. If the secondary coil is part of a complete circuit an alternating current is produced.

- The coils of wire are insulated so that current does not short across either the iron core or adjacent turns of wire, but flows around the whole coil. The core is made of iron so it is easily magnetised.

- Transformers are used in the **National Grid**.
  - A **step-up transformer** makes the pd across the secondary coil greater than the pd across the primary coil. Its secondary coil has more turns than its primary coil.
  - A **step-down transformer** makes the pd across the secondary coil less than the pd across the primary coil. Its secondary coil has fewer turns than its primary coil.

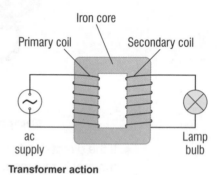

Transformer action

1 *Why is the core of a transformer made of iron not copper?*

- A **switch mode transformer** has a ferrite core. Compared with a traditional transformer, a switch mode transformer:
  - operates at a much higher frequency
  - is lighter and smaller
  - uses very little power when there is no device connected across its output terminals.

2 *What happens if a 1.5 V cell is used as the supply for the primary coil?*

Study tip

Remember there is no current in the iron core, just a magnetic field.

Key words: transformer, National Grid, step-up transformer, step-down transformer, switch mode transformer

3.5 Transformers in action

- Transformers are used to step potential differences up or down.

- $\dfrac{V_p}{V_s} = \dfrac{n_p}{n_s}$

- For a step-down transformer, n_s is less than n_p

- For a step-up transformer, n_s is greater than n_p.

- For a 100% efficient transformer:

  $V_p \times I_p = V_s \times I_s$

Maths skills

A step-up transformer is used to change a pd of 12V to a pd of 120V. If there are 50 turns on the primary coil, how many turns are there on the secondary coil?

$V_p = 12\,\text{V}$

$V_s = 120\,\text{V}$

$n_p = 50$

$$\frac{V_p}{V_s} = \frac{n_p}{n_s}$$

$$n_s = \frac{n_p V_s}{V_p}$$

$$n_s = \frac{50 \times 120\,\text{V}}{12\,\text{V}}$$

$$n_s = 500$$

There are 500 turns on the secondary coil.

- The National Grid uses transformers to step-up the pd from power stations.
- This is because the higher the pd at which electrical energy is transmitted across the Grid, the smaller the energy wasted in the cables.
- Step-down transformers are used to reduce the pd so that it is safe to be used by consumers.
- The pd across, and the number of turns on, the primary and secondary coils are related by the equation:

$$\frac{V_p}{V_s} = \frac{n_p}{n_s}$$

Where:
V_p is the pd across the primary coil in volts, V
V_s is the pd across the secondary coil in volts, V
n_p is the number of turns on the primary coil
n_s is the number of turns on the secondary coil.

- A step-down transformer has fewer turns on the secondary coil than on the primary coil.
- A step-up transformer has more turns on the secondary coil than on the primary coil.

> **1** *Why is a transformer used to step-up the pd from a power station?*

- Transformers are almost 100% efficient. For 100% efficiency:

$$V_p \times I_p = V_s \times I_s$$

Where:
I_p is the current in the primary in amperes, A,
I_s is the current in the secondary in amperes, A.

> **2** *A transformer has 100 turns on the primary coil and 400 turns on the secondary coil. The pd across the primary coil is 2V. What is the pd across the secondary coil?*

Bump up your grade

Make sure you can rearrange the transformer equation.

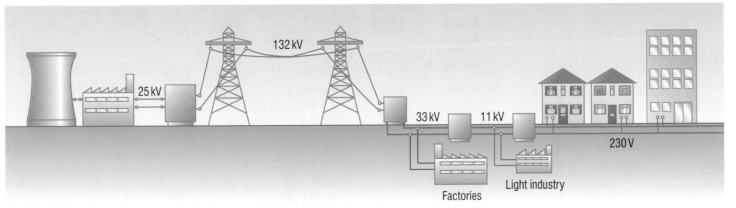

The grid system

3.6 A physics case study

- Applications of physics are used in hospitals for both diagnosis and therapy.
- An ECG or electrocardiogram is used to measure the potential differences generated by the heart.
- Electronic devices are used to measure the blood pressure.
- Digital thermometers are used to measure temperature.
- An endoscope containing bundles of fibre optics is used to look inside the body without making large incisions.
- X-rays are used to take pictures of suspected broken bones.
- CT scanners are used to build up digital pictures of a cross-section through the body.
- MR scanners use radio waves to produce detailed digital pictures of the body.

1 *Which part of the electromagnetic spectrum is used in a CT scanner?*

> **Study tip**
>
> In the exam you will be expected to apply your physics knowledge to unfamiliar situations.

2 *Why do broken bones show up on X-rays?*

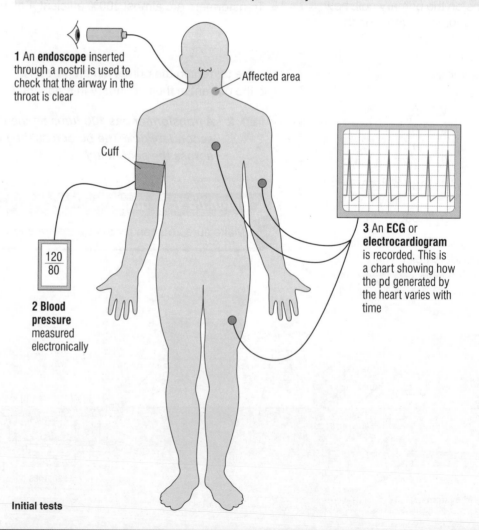

1 An **endoscope** inserted through a nostril is used to check that the airway in the throat is clear

Affected area

Cuff

120
80

2 Blood pressure measured electronically

3 An **ECG** or **electrocardiogram** is recorded. This is a chart showing how the pd generated by the heart varies with time

Initial tests

1 Why must insulated wire be used to make an electromagnet?

2 What is a magnetic field line?

3 What does the thumb represent in Fleming's left-hand rule?

4 How can the size of the force in a motor be increased?

5 What happens if a magnet is held at rest in a coil and the coil is pulled off it?

6 How can the size of an induced pd be increased?

7 What does a transformer consist of?

8 How does the frequency of operation of a switch mode transformer compare with a traditional transformer?

9 Where is a step-down transformer used in the National Grid?

10 A transformer is used to step-down a 240 V supply to 12 V. There are 100 turns on the primary coil. How many turns are there on the secondary coil?

11 A transformer is used to step-down a 240 V supply to 12 V. If the current in the primary coil is 0.1 A, what is the current in the secondary coil?

12 What is the core of a switch mode transformer made from?

Chapter checklist			✓	✓	✓
Tick when you have:			Electromagnets		
reviewed it after your lesson	✓ ☐ ☐		The motor effect		
revised once – some questions right	✓ ✓ ☐		Electromagnetic induction		
revised twice – all questions right	✓ ✓ ✓		Transformers		
Move on to another topic when you have all three ticks			Transformers in action		
			A physics case study		

1   The diagram shows a seesaw. The centre of mass of the seesaw is at its centre.

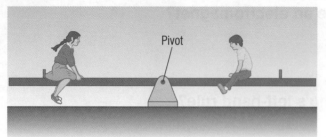

Pivot

a   What is meant by 'centre of mass of the seesaw'?   *(1 mark)*

b   A child of weight 460 N sits 2.0 m from the pivot. A second child sits on the other side of the pivot and the seesaw becomes balanced.

   i   Why must the second child sit on the other side of the pivot?   *(2 marks)*

   ii   The weight of the second child is 575 N. How far away from the pivot must the second child sit for the seesaw to become balanced?

        Show clearly how you work out your answer and give the unit.   *(4 marks)*

c   Explain what would happen if the children were to sit one at each end of the seesaw.   *(2 marks)*

Study tip

Qu 1bii There are four marks for this calculation. Make sure that you show your working step by step so that if you make a mistake you may still get some credit.

2   *In this question you will be assessed on using good English, organising information clearly and using specialist terms where appropriate.*

The diagram shows a moving coil loudspeaker. The loudspeaker contains a moveable coil attached to a diaphragm. The diaphragm fits loosely over a cylindrical permanent magnet. An amplifier produces a varying, alternating current in the coil.

Explain how the loudspeaker makes use of the motor effect to produce a sound wave.
                                                                         (6 marks)

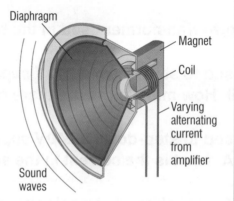

Diaphragm
Magnet
Coil
Varying alternating current from amplifier
Sound waves

Study tip

There are six marks available for this question. To gain all of them your explanation must be in a logical order. Before you write anything stop and think about the points you want to make and the order you want to put them in.

3   The diagram shows a human eye.

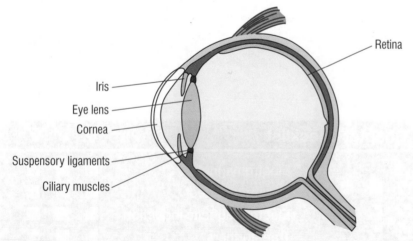

Retina
Iris
Eye lens
Cornea
Suspensory ligaments
Ciliary muscles

Light enters the eye and is focused on the retina.

a   Explain how the amount of light entering the eye is controlled.   *(2 marks)*

b   Explain how the light is brought to focus on the retina.   *(2 marks)*

4 The diagram shows the basic structure of a step-down transformer.

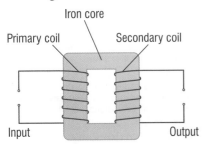

a Explain how the transformer works. *(5 marks)*

b The transformer is used to change the 230 V mains supply to a 12 V supply to operate a model train.

  i There are 30 turns on the secondary coil. Calculate the number of turns on the primary coil.
Write down the equation you use and show clearly how you work out your answer. *(3 marks)*

  ii The current drawn from the mains electricity supply by the transformer is 0.048 A. Calculate the current through the model train.
Assume that the transformer is 100% efficient.
Write down the equation you use. Show clearly how you work out your answer and give the unit. *(2 marks)*

5 A doctor wants to look inside a patient's stomach, without operating on him.

a The doctor uses an endoscope. The endoscope contains bundles of optical fibres. Explain how visible light and bundles of optical fibres are used in the endoscope to look inside the patient's stomach. *(3 marks)*

b The tube below the stomach is called the small intestine. The doctor wants to take an X-ray picture of the small intestine. Before the picture is taken, the patient is given a drink containing barium, a substance that absorbs X-rays.

  i Explain why a normal X-ray will not allow the doctor to see the small intestine. *(2 marks)*

  ii Explain why giving the patient barium allows the doctor to see the small intestine on an X-ray picture. *(2 marks)*

  iii During this procedure the radiographer stands behind a lead screen. Explain why the radiographer needs to stand behind a lead screen but the patient does not. *(4 marks)*

Answers

B3 Answers

1 Exchange of materials

1.1
1 Partially permeable

1.2
1 The particles are being absorbed against a concentration gradient.

1.3
1 Glucose is used in the process of respiration to release energy.
2 During sweating

1.4
1 Alveoli

1.5
1 Intercostal muscles *and* diaphragm

1.6
1 One from: It is much smaller/the person does not have to lie in a metal case/some can be used in the home/they can be controlled by computers to give the correct amount of air.

1.7
1 Two of: large surface area, thin walls, good blood supply

1.8
1 To allow gases to diffuse in and out of the leaf.

1.9
1 Diffusion

Answers to end of chapter questions

1 Diffusion refers to the movement of any particles from a region of high concentration to a region of lower concentration. Osmosis only refers to the diffusion of water.
2 Active transport is used by roots to absorb mineral ions (or by kidney tubules to reabsorb glucose).
3 Particles move from a region of lower concentration to a region of higher concentration.
4 Lungs: alveoli, small intestine: villi
5 Large surface area, thin walls or short diffusion path, good transport system – the blood supply in animals
6 The rib muscles contract, the diaphragm contracts. The ribcage moves up and out, the diaphragm flattens. This increases the volume inside the chest and decreases the pressure, so air is forced into the lungs.
7 Patients who are paralysed and cannot breathe for themselves.
8 By the root hair cells which have projections called root hairs to absorb water and mineral ions from the soil.
9 The plant is photosynthesising at a faster rate and needs to take in more carbon dioxide. The stomata will be open as wide as possible.
10 The air already contains a lot of water vapour, so there is not a steep concentration gradient, therefore diffusion of water vapour from the leaf is much reduced.

2 Transporting materials

2.1
1 Valves prevent the backflow of blood.

2.2
1 Arteries have thicker walls than veins. Veins have valves along their length, arteries do not have valves along their length.
2 To open up blood vessels which have been narrowed by fatty deposits, particularly the coronary arteries.

2.3
1 Carbon dioxide
2 They help blood to clot at the site of a wound.

2.4
1 In an emergency to increase blood volume when a lot of blood has been lost by the patient
2 They do not need to match the person's tissue, there is no need for immunosuppressant drugs and they are available when there are not enough donors.

2.5
1 Xylem and phloem

Answers to end of chapter questions

1 Capillaries
2 Haemoglobin
3 Arteries
4 Valves prevent backflow.
5 White blood cells are part of the body's defence system.
6 a Xylem transports water and mineral ions through the plant.
   b Phloem transports sugars.
7 Blood must be circulated to the lungs to become oxygenated. Oxygenated blood must be returned to the heart to be pumped around the body.
8 The stent keeps the coronary artery open so that oxygenated blood can flow to the heart muscle.
9 There are not enough heart donors for the number of patients who have heart disease and require a new heart.
10 The artificial heart would need two pumps – one to pump blood to the lungs and the other to pump blood to the rest of the body. It would also need valves to prevent backflow. It should be made of materials which do not cause an immune reaction.

3 Keeping internal conditions constant

3.1
1 Urea is made in the liver.

3.2
1 Filtration of the blood and reabsorption of substances needed by the body.

3.3
1 Dialysis removes urea from the blood.

3.4
1 Proteins on the surface of cells
2 The patient does not have to be attached to a machine every few days.

3.5
1 In the thermoregulatory centre in the brain

3.6
2 The water in the sweat evaporates. Energy from the skin is used to turn the water into water vapour so the skin cools. [H]

3.6
1 Donors are usually people who have been killed in an accident. It is distressing for families to have to make decisions just after they have lost a close relative.

3.7
1 Insulin

3.8
1 Any two from: pancreas transplants; transplanting pancreas cells; using embryonic stem cells to produce insulin secreting cells; using adult stem cells from diabetic patients; genetically engineering pancreas cells to make them work properly.

Answers to end of chapter questions

1 Urea is a toxic (poisonous) substance.
2 Glucose is needed by the body (for respiration).
3 Active transport
4 Partially permeable membrane
5 Immunosuppressant drugs are given to the patient.
6 The surfaces of cells have proteins called antigens attached to them. Antibodies recognise foreign antigens and attack them. If the donor kidney has similar antigens to the recipient the antibodies will not attack them.
7 The enzymes do not work properly – too hot they denature – too cold they work too slowly and respiration is too slow to release energy.
8 Glucagon causes glycogen to be turned into glucose which enters the blood and raises the blood glucose level. [H]
9 The insulin would be digested by proteases in the stomach and would not work.
10 Eating carbohydrates/sugar increases the level of sugars/glucose in the blood. Reducing carbohydrates helps to keep blood sugar levels down. Increased exercise means more glucose is used by the cells and the blood glucose falls – less insulin is needed by the diabetic. A combination of too much carbohydrate and too little exercise would mean more insulin needed in the injection.

4 How humans can affect the environment

4.1
1 Land has to be cleared, which destroys their habitats.

4.2
1 Toxins leak out into the surrounding soil and into underground water.

4.3
1 Damage to trees means birds lose food and habitats.

4.4
1 Photosynthesis
2 Decay by microorganisms

4.5
1 Carbon dioxide is removed from the air by plants in photosynthesis and by dissolving in oceans, rivers and lakes.
2 If the climate gets warmer the ice will melt.

4.6
1 By distillation

▶ 4.7
1 Methane

▶ 4.8
1 To prevent them using energy (wasting energy) for movement, and to maintain their body temperature. The energy is used for growing (turning into meat).

▶ 4.9
1 The population could become extinct if the young fish don't grow and reproduce.
2 Mycoprotein

▶ 4.10
1 To build reservoirs to store water for human consumption

Answers to end of chapter questions

1 Chemicals used to kill weeds
2 Sulfur dioxide
3 The clouds of gas are blown by the wind.
4 The fertiliser contains nitrates. Large amounts of nitrate enter the rivers (which alters the balance of chemicals in the water).
5 Methane is produced by cattle and from growing rice.
6 Burning the trees releases carbon dioxide. When the trees decay, carbon dioxide is released due to respiration of microorganisms.
7 Plants are grown which use carbon dioxide from the atmosphere during photosynthesis. When the products from the plants are fermented, ethanol is produced. Burning ethanol releases the same amount of carbon dioxide back into the atmosphere as was removed by photosynthesis.
8 In aerobic conditions, the bacteria would produce carbon dioxide and not methane.
9 Oxygen levels, pH, temperature
10 Human activities cause an increase in greenhouse gases. Energy is radiated back to Earth from the greenhouse gases in the atmosphere and warms the planet.

Answers to practice questions

1 a Transports sugar to storage organs *(1 mark)*
  b Transports water up the stem *(1 mark)*
2 a A [If B given, no marks throughout; if unspecified plus two good reasons, 1 mark] High(er) pressure in A [Allow opposite for B, do not accept 'zero pressure' for B] Pulse/described in A [Accept fluctuates/'changes', allow reference to beats/beating, ignore reference to artery pumping] *(2 marks)*
  b i 17 *(1 mark)*
    ii 68 *(1 mark)* [Accept correct answer from candidate's bi × 4]
  c i Oxygen/oxygenated blood *(1 mark)* [Allow adrenaline, ignore air] glucose/sugar *(1 mark)* [Extra wrong answer cancels, e.g. sucrose/starch/glycogen/glucagons/water, allow fructose as an alternative to glucose, ignore energy, ignore food]
    ii Carbon dioxide/CO_2/lactic acid *(1 mark)* [Allow CO2/CO^2, ignore water] [Answers to this AQA examination question have not been supplied or approved by AQA]
3 a 10 times magnification *(2 marks)* [distance between arrows/0.5 = 1 mark, if unit given minus 1]
  b i Water moves into the amoeba from a dilute solution/region of high water concentration *(1 mark)* to a more concentrated solution *(1 mark)* across a partially permeable membrane *(1 mark)* otherwise the amoeba would swell up and

burst *(1 mark)*
    ii Sea water contains salt/more concentrated than fresh water/similar concentration to cytoplasm *(1 mark)* water would move in and out at similar rate/idea of equilibrium/no gradient *(1 mark)*
4 a i Any one from: Compete (with *Fusarium*) for food/oxygen or reduce yield of *Fusarium*. Make toxic waste products **or** they might cause disease/pathogenic **or** harmful to people/*Fusarium*. *(1 mark)* [Do **not** allow harmful unqualified.]
    ii Any two from: Steam/heat treat/sterilise fermenter (before use). [Not just clean, allow sterilisation unqualified for 1 mark] Steam/heat treat/sterilise glucose/minerals/nutrients/water (before use). [Not just use pure glucose)] Filter/sterilise air intake. Check there are no leaks. *(2 marks)*
  b Any three from: Beef is best or beef is better than mycoprotein*. Mycoprotein is <u>mainly</u> better than wheat*. There is more phenylalanine in wheat than in mycoprotein*. [*Allow equivalent numerical statements.] But no information given on other amino acids/costs/foods. *(3 marks)* Overall conclusion: Statement is incorrect. or It would be the best source for vegetarians. or For given amino acids, beef is the best source. or Three foods provide insufficient data to draw a valid conclusion. *(1 mark)* [Answers to this AQA examination question have not been supplied or approved by AQA]
5 a Any two of: may cause big changes in the Earth's climate may cause a rise in sea level may reduce biodiversity may cause changes in migration patterns, e.g. in birds which may result in changes in the distribution of species *(2 marks)*
  b Marks awarded for this answer will be determined by the Quality of Written Communication (QWC) as well as the standard of the scientific response.

There is a clear, balanced and detailed description of a variety of ways in which human activities affect the levels of both carbon dioxide and methane, with points from the examples below. The answer shows almost faultless spelling, punctuation and grammar. It is coherent and in an organised, logical sequence. It contains a range of appropriate or relevant specialist terms used accurately. *(5–6 marks)*
There is a description of a range of the ways in which human activities affect the levels of both carbon dioxide and methane, with points from the examples below. There are some errors in spelling, punctuation and grammar. The answer has some structure and organisation. The use of specialist terms has been attempted, but not always accurately. *(3–4 marks)*
There is a brief description of at least two ways in which human activities affect the levels of either or both carbon dioxide and methane, which has little clarity and detail. The spelling, punctuation and grammar are

very weak. The answer is poorly organised with almost no specialist terms and/or their use demonstrating a general lack of understanding of their meaning. *(1–2 marks)*
No relevant content. *(0 marks)*

Examples of biology points made in the response:
- (for 3–6 marks both carbon dioxide and methane must be mentioned)
- (large scale) deforestation
- to clear areas for farming/agriculture
- (deforestation) reduces up take of carbon dioxide in the atmosphere
- (deforestation) releases carbon dioxide into the atmosphere due to burning
- (deforestation) releases/increases carbon dioxide into the atmosphere due to decay by/respiration of microorganisms
- destruction of peat bogs releases carbon dioxide
- peat used for compost/fuel
- increased need for food
- rice growing releases methane
- increase in cattle releases methane.

C3 Answers

1 The periodic table

▶ 1.1
1 Protons (and electrons) had not been discovered or they did not know about atomic structure.
2 The elements fitted better into the groups (because he had left gaps).

▶ 1.2
1 They have the same number of electrons in the highest occupied energy level or outer shell.
2 When metals react their atoms lose electrons, as the atoms get larger there are more occupied energy levels and the electrons in the highest occupied level or outer shell are less strongly attracted by the nucleus and so are lost more easily.

▶ 1.3
1 They react with water to produce alkalis or their hydroxides are alkalis.
2 Potassium bromide, KBr
3 Lithium atoms are smaller than sodium atoms, they have fewer occupied energy levels (Li 2,1; Na 2,8,1), their outer electron is more strongly attracted by the nucleus, and so is less easily lost when they react (to form an ion).

▶ 1.4
1 They are strong, hard, react only slowly (or not at all) with oxygen/air and water (at ordinary temperatures).
2 Higher melting points, stronger, harder, denser, less reactive (with oxygen and water), ions have different charges, coloured compounds, catalysts.

▶ 1.5
1 They have small molecules (made of pairs of atoms).
2 Add chlorine (water) to a solution of potassium bromide (or other soluble bromide), bromine will be displaced.

Answers to end of chapter questions

1 When the elements are arranged in order of atomic weights, every eighth element has similar properties.
2 The elements fitted the groups better or the elements within a group all had similar properties, it allowed for undiscovered elements or left gaps for undiscovered elements.
3 They have the same number of electrons in the

highest occupied energy level or outer shell (same number of outer electrons).

4 a lithium + water → lithium hydroxide + hydrogen

  b Three from: lithium floats, moves around the surface, gradually disappears, bubbles (of gas) or fizzes.

  c Add (universal) indicator, goes purple or blue (or correct alkaline colour for named indicator).

  d Sodium reacts faster or melts (lithium does not melt).

5 Physical: (three from) high melting point (and/ or boiling point), hard, strong, high density, malleable, ductile, good conductor (of heat and electricity), can be made into alloys. Chemical: (three from) unreactive or reacts slowly with oxygen (air) and/or water, forms positive ions/ionic compounds, forms ions with different charges, coloured compounds, catalyst.

6 They increase.

7 NaBr; colourless or white, crystals or solid; dissolves in water, forms a colourless solution.

8 Covalently bonded, small molecules (has weak forces between its molecules).

9 a From colourless to orange/yellow/brown.

  b chlorine + potassium bromide → potassium chloride + bromine

  c $Cl_2 + 2KBr \rightarrow 2KCl + Br_2$ [H]

10 $2Fe + 3Cl_2 \rightarrow 2FeCl_3$ [H]

11 a A sodium atom has more occupied energy levels/shells, so its outer electron is further from and less strongly held by the nucleus, and so can be more easily lost when it reacts.

  b A fluorine atom has fewer occupied energy levels/shells, so its nucleus has a greater attraction for electrons in the highest occupied energy level/outer shell, so it attracts electrons more readily when it reacts. [H]

2 Water

2.1

1 Soft water lathers easily with only a little soap, hard water uses more soap and produces scum.

2 Scum is formed when soap reacts with compounds in hard water. Scale is produced when temporary hard water is heated or boiled.

3 Hard water contains calcium compounds that are good for health, for the development of teeth and bones, and to reduce the risk of heart disease.

2.2

1 Water that is softened by heating or boiling, (and forms scale when heated).

2 $Ca^{2+}(aq) + CO_3^{2-}(aq) \rightarrow CaCO_3(s)$ [H]

3 Ion-exchange resin removes the calcium and magnesium ions from hard water and replaces them (exchanges them) for sodium ions or hydrogen ions (that do not react with soap).

2.3

1 Sedimentation and/or filtration to remove solids, killing microbes (disinfecting/sterilising) using chlorine (or other methods, e.g. ozone, ultraviolet).

2 It requires a large amount of energy and so it would be very expensive and/or tap water does not need to be pure but should be free from harmful substances.

2.4

1 Advantage: (one of) uses less soap, does not produce scum, reduces the effects on heating systems. Disadvantage (one of): reduces health benefits, costs money.

2 No chlorine: increased health risks, more (water-borne) spreading of disease. No fluoride: more tooth decay or poorer dental health.

Answers to end of chapter questions

1 Soap reacts with calcium and/or magnesium ions/compounds to produce insoluble solids.

2 Scale (insoluble solids of calcium carbonate and magnesium carbonate)

3 Health benefit, e.g. stronger teeth and bones or reduced risk of heart attacks

4 Temporary hard water is softened or produces scale when heated/boiled, permanent hard water is not softened or does not produce scale when heated/boiled.

5 Washing soda is sodium carbonate which is soluble, calcium ions and/or magnesium ions in the water react with carbonate ions to produce precipitates/insoluble calcium carbonate and/or magnesium carbonate, thus removing the ions from the water so they cannot react with soap.

6 It replaces calcium ions with sodium ions that can have negative health effects or increase the risk of heart disease.

7 Insoluble solids

8 To kill microbes/bacteria (to disinfect or sterilise the water).

9 No need for drinking water to be pure water or drinking water just needs to be free from harmful substances.

10 To reduce the amount of tooth decay or to improve dental health.

11 Hydrogencarbonate ions decompose when heated to produce carbonate ions, water and carbon dioxide:
$2HCO_3^-(aq) \rightarrow CO_3^{2-}(aq) + H_2O(l) + CO_2(g)$
Carbonate ions react with calcium ions and/or magnesium ions in the water to produce precipitates of calcium carbonate and/or magnesium carbonate that are deposited as scale:
$Ca^{2+}(aq) + CO_3^{2-}(aq) \rightarrow CaCO_3(s)$ or
$Mg^{2+}(aq) + CO_3^{2-}(aq) \rightarrow MgCO_3(s)$ [H]

3 Energy calculations

3.1

1 36 400 J or 36.4 kJ

2 Fuel B (A releases 2620.8 kJ/mol and B releases 4058.4 kJ/mol)

3.2

1 7560 J or 7.56 kJ

2 142.8 kJ/mol of iron

3.3

1 Energy level diagram similar to figure but with the products at a higher energy level than the reactants.

2 Energy level diagram similar to figure but with the products at a higher level than the reactants, the reaction pathway rising above the products level, and the activation energy labelled with an arrow pointing upwards from the reactants level to the top of the pathway curve.

3 Energy level diagram similar to figure but with the products at a higher level than the reactants. The catalysed pathway should be labelled and its peak should be below the peak for the uncatalysed pathway (i.e. has a lower activation energy).

3.4

1 Bonds broken: $(2 \times C–C) + (8 \times C–H) + (5 \times O=O)$, energy needed = 6488 kJ,
Bonds made: $6 \times C=O + 8 \times H–O = 8542$ kJ
Energy change of reaction = 2054 kJ/mol [H]

3.5

1 Fossil fuels are non-renewable or a limited resource, fossil fuels cause pollution or release carbon dioxide that causes global warming, hydrogen produces only water or does not produce carbon dioxide or other pollutants, hydrogen releases a large amount of energy per gram, hydrogen can be produced from renewable sources.

Answers to end of chapter questions

1 Energy losses or it is not only the water that is heated.

2 The energy level of the reactants is below the level of the products.

3 Diagram similar to third figure in C3 3.3, with reactants on line above products, reaction pathway shown as curved line reaching maximum above reactants, activation energy shown as an arrow from reactants to maximum of pathway and energy change of reaction shown as arrow pointing downwards from reactants to products.

4 12600 J or 12.6 kJ (or 25.2 kJ/g)

5 75.6 kJ/mol

6 58.8 kJ/mol

7 Diagram similar to third figure in C3 3.3, with reactants on line above products and two reaction pathways, the one with the lower maximum labelled 'with catalyst'.

8 Three from: supply, storage, safety, performance, convenience, price/cost of vehicles

9 542 kJ/mol [H]

10 1318 kJ/mol [H]

4 Analysis and synthesis

4.1

1 Lithium (Li^+) and calcium (Ca^{2+})

2 Calcium (Ca^{2+}) and magnesium (Mg^{2+})

4.2

1 Hydrochloric acid contains chloride ions (Cl^-) and sulfuric acid contains sulfate ions (SO_4^{2-}) which both give precipitates with silver nitrate solution.

4.3

1 A pipette measures a fixed volume, has a single graduation, and has no tap; a burette measures different volumes, has graduations, has a tap.

2 To show when the end-point is reached or when the reaction is complete or when the acid and alkali have reacted exactly.

4.4

1 12 g/dm³, 0.5 mol/dm³ [H]

2 0.17 mol/dm³ [H]

4.5

1 When we need to know how much or the quantity of a substance in a sample.

2 DNA results/profile are different for every individual.

4.6

1 When the rates of the forward and reverse reactions of a reversible reaction are equal or when the amounts of reactants and products in a reversible reaction are constant. [H]

2 So that more of the reactants react or so more SO_3 is produced. [H]

4.7

1 An increase in pressure

2 A decrease in temperature

4.8

1 nitrogen + hydrogen ⇌ ammonia

2 Unreacted gases are recycled.

4.9

1 More energy is needed and stronger reaction vessels and pipes are needed which both cost more. **[H]**

2 The reaction would be too slow (rate decreased and catalyst will not work) **[H]**

Answers to end of chapter questions

1 Fe^{2+}/iron(II) ions and SO_4^{2-}/sulfate ions

2 Potassium iodide (KI)

3 Copper(II) carbonate ($CuCO_3$)

4 A pipette is used to measure a fixed volume (e.g. $25\,cm^3$) of (alkali) solution (into a conical flask). A burette is used to add (acid) solution (gradually) to the flask until the end-point of the reaction is reached and to measure the volume that has been added.

5 a Air (nitrogen), natural gas (hydrogen)

   b About 200 atmospheres pressure, about 450°C, iron catalyst.

   c The gases are cooled (as they leave the reactor) and ammonia condenses (liquefies) but nitrogen and hydrogen remain as gases.

6 $0.05\,mol/dm^3$ **[H]**

7 By increasing the temperature, by removing carbon dioxide or allowing it to escape. **[H]**

5 Organic chemistry

5.1

1 methane, ethane and propane

2 $CH_3CH_2CH_2OH$

3 Displayed formula of HCOOH

5.2

1 It is a solvent (it mixes with water, it evaporates easily)

2 One of: apply a flame – ethanol burns in air: add sodium – ethanol reacts more slowly (also sinks in ethanol): add (acidified) potassium dichromate (or other oxidising agent) and heat – colour change or smell of vinegar

5.3

1 They produce $H^+(aq)$, hydrogen ions in aqueous solutions.

2 They have distinctive smells/tastes that are fruity/pleasant.

3 $CH_3CH_2COOH(aq) \rightleftharpoons CH_3CH_2COO^-(aq) + H^+(aq)$ **[H]**

5.4

1 Advantages and disadvantages may change over time as resources are used up or there are new developments or new research.

Answers to end of chapter questions

1 Methanol, ethanol and propanol, CH_3OH, CH_3CH_2OH, $CH_3CH_2CH_2OH$

2 Propanoic acid, CH_3CH_2COOH

3
Ethyl ethanoate

4 Dissolves many substances, mixes with water, evaporates easily (volatile).

5 It reacts, fizzes/effervesces, produces hydrogen (slower than with water, sodium sinks)

6 Microbes in the air, caused oxidation of ethanol, produced ethanoic acid.

7 Reacts with carbonates to produce carbon dioxide (and a salt and water) or reacts with metals to produce hydrogen and a salt, or reacts with bases/alkalis to produce a salt and water.

8 Ethyl ethanoate, (sulfuric) acid catalyst, heat the mixture.

9 It is an ester and so has a fruity smell/flavour.

10 One advantage, e.g. can be produced from renewable resources, does not release locked-up carbon or is carbon neutral; one disadvantage, e.g. need food crops/land for production.

11 Test with universal indicator or pH meter/probe, ethanoic acid has higher pH: or add a metal or carbonate, ethanoic acid reacts slower. **[H]**

12 $CH_3CH_2CH_2OH + 4.5O_2 \rightarrow 3CO_2 + 4H_2O$ or $2CH_3CH_2CH_2OH + 9O_2 \rightarrow 6CO_2 + 8H_2O$ **[H]**

Answers to Practice questions

1 a chlorination, filtration *(2 marks)*

   b calcium, magnesium *(2 marks)*

   c add soap solution and shake, scum forms or does not lather *(2 marks)*

   d i temporary hard water *(1 mark)*

      ii reduces efficiency of water heating systems, blocks pipes *(1 mark)*

   e add sodium carbonate (washing soda), use ion exchange column/resin *(2 marks)*

   f Two from: too expensive, no need to soften all water supplied, hard water is better for health, water only needs to be safe to drink, choice left to consumer. *(2 marks)*

   g Four from: Temporary hard water contains hydrogencarbonate ions (HCO_3^-), these decompose on heating producing carbonate ions, $2HCO_3^- \rightarrow CO_2 + CO_3^{2-} + H_2O$, carbonate ions react with calcium and magnesium ions to form precipitates (removing the ions from solution), $Ca^{2+} + CO_3^{2-} \rightarrow CaCO_3$ or $Mg^{2+} + CO_3^{2-} \rightarrow MgCO_3$. **[H]** *(4 marks)*

2 a alcohols *(1 mark)*

   b CH_3CH_2OH *(1 mark)*

   c –OH *(1 mark)*

   d Marks awarded for this answer will be determined by the Quality of Written Communication (QWC) as well as the standard of the scientific response.

   There is a clear and detailed description showing the detailed understanding of the method. Knowledge of accurate information appropriately contextualised. The answer shows almost faultless spelling, punctuation and grammar. It is coherent and in an organised, logical sequence. It contains a range of appropriate and relevant specialist terms used accurately. *(5–6 marks)*
   There is some description showing a clear understanding of the method. There are some errors in spelling, punctuation and grammar. The answer has some structure and organisation, use of specialist terms has been attempted but not always accurately. *(3–4 marks)*
   There is a brief description of the method. The spelling, punctuation and grammar are very weak. The answer is poorly organised, with almost no specialist terms and their use demonstrates a general lack of understanding of their meaning. *(1–2 marks)*
   No relevant content. *(0 marks)*
   Examples of chemistry points made in the candidate's response:
   • Measure appropriate volume of water in measuring cylinder and put into into calorimeter.
   • Measure the initial temperature.
   • Measure the mass of the spirit burner.
   • Ignite spirit burner and place under calorimeter. Allow to burn until suitable

   temp rise or for a few minutes. Put out flame. Stir water.
   • Record the highest steady temperature.
   • Re-weigh spirit burner.
   • Calculate temperature change.
   • Calculate mass of compound burned.
   • Assume volume water (cm^3) = mass of water (g). Calculate energy change for 1 g (or 1 mole) of each fuel to compare.
   • Evidence of high level work e.g. appropriate quantities, use of precise balance, appropriate size of measuring cylinder, attempt to reduce draughts, e.g. use of shields, using same sized flame, same distance from calorimeter base.

   e One mark for each of the following:
   Energy level diagram with reactants (labelled) above level of products
   curve showing energy change during reaction that peaks above reactants
   activation energy labelled as arrow from level with reactants to maximum of curve
   energy change of reaction labelled as arrow from level with reactants to level with products. *(4 marks)*

3 a The air *(1 mark)*

   b e.g. natural gas, crude oil/naphtha/product of cracking hydrocarbons, (allow electrolysis of water or brine – not usually used but are possible sources). *(1 mark)*

   c Catalyst or to speed up the reaction. *(1 mark)*

   d recycled or returned to the reactor *(1 mark)*

   e cooled or temperature decreased *(1 mark)*

   f reversible reaction *(1 mark)*

   g There are four molecules (moles) of reactant gases and two molecules of product gases or fewer molecules of gas in products than reactants (1) so high pressure favours forward reaction. **[H]** *(2 marks)*

   h The forward reaction is exothermic or reverse reaction is endothermic (1) high temperature favours the endothermic reaction or the reverse reaction or low temperature favours the exothermic reaction (1) **[H]** *(2 marks)*

   i The rate of reaction is higher at higher temperatures (1)
   so ammonia is produced quickly (1)
   or
   converse argument gains 2 marks or the catalyst only works at high temperatures gains 1 mark. **[H]** *(2 marks)*

4 a add hydrochloric acid (or other named acid) carbonate effervesces/fizzes or produces carbon dioxide gas, nitrate has no reaction. *(2 marks)*

   b Add dilute nitric acid and silver nitrate solution white precipitate with chloride, yellow precipitate with iodide. *(2 marks)*

   c Flame test
   calcium chloride gives red flame colour, magnesium chloride gives no colour *(2 marks)*

   d Add sodium hydroxide solution
   green precipitate with iron(II) sulfate and brown precipitate with iron(III) sulfate. *(2 marks)*

   e Test pH of aqueous solution using named indicator or pH meter/probe
   ethanol solution pH 7 or appropriate neutral colour of indicator (e.g. UI green) and ethanoic acid pH < 7 or appropriate acid colour of indicator (e.g. UI red) **[H]** *(2 marks)*

5 a Bonds broken: $4 \times C–C + 12 \times C–H + 8 \times O=O = 10328$
   Bonds made: $10 \times C=O + 12 \times C–H = 13618$
   Energy change = 3290 kJ *(5 marks)*

   b Energy released when bonds made (in products) is greater than energy needed to break bonds (in reactants) *(1 mark)*

P3 Answers

1 Medical applications of physics

1.1
1 To shield them, the lead absorbs X-rays.

1.2
1 20 000 Hz

1.3
1 The change in direction of a wave as it crosses from one transparent material to another.
2 It refracts away from the normal.

1.4
1 42°   [H]
2 A very thin, flexible glass fibre

1.5
1 The point where rays of light parallel to the principal axis are brought to a focus
2 By placing an object between the lens and the principal focus

1.6
1 Rays drawn on a diagram to determine the position and nature of an image
2 Real

1.7
1 Cornea and lens
2 4.0 d

1.8
1 A diverging lens
2 The eyeball is too short, or eye lens too weak.

Answers to end of chapter questions
1 To monitor their exposure to X-rays
2 A device that uses X-rays to produce digital images of a cross-section through the body
3 A sound wave with a frequency greater than the upper limit for human hearing, i.e. 20 000 Hz
4 X-rays are ionising, so might harm the baby.
5 9.7°   [H]
6 It passes into the block without changing direction.
7 It will be refracted at an angle of 90°, along the boundary between the glass and the air.
8 The point that parallel rays of light appear to have diverged from after they have passed through the lens.
9 Virtual, upright, smaller than the object
10 The muscles that control the thickness of the lens in the eye
11 6.25 d
12 1.5 cm

2 Using physics to make things work

2.1
1 12 Nm
2 40 N   [H]

2.2
1 The point where the mass of the object can be thought to be concentrated
2 Along the axis of symmetry

2.3
1 Because their perpendicular distance to the pivot is zero
2 1 m   [H]

2.4
1 The bags raise the centre of mass so the pushchair will not have to tilt so far before the line of action of the weight moves outside the base.

2 So they are likely to topple over when hit by a bowling ball

2.5
1 7500 Pa
2 Liquids are virtually incompressible, and the pressure in a liquid is transmitted equally in all directions.

2.6
1 Tension in the string
2 The conker will fly off in a straight line at a tangent to the circle.

2.7
1 The length of the string
2 0.1 s

Answers to end of chapter questions
1 The turning effect of a force
2 A crowbar allows you to apply the same force at a greater distance from the pivot, giving a bigger moment.
3 a   b   c

4 It is continuously changing direction, so it is continuously changing velocity.
5 7.5 Nm
6 Directly below the point of suspension
7 2500 Pa
8 An acceleration towards the centre of a circle
9 It transfers energy doing work against friction at the pivot and air resistance.
10 50 Hz
11 The distance from the highest point on one side of the oscillation to the highest point on the other side.
12 When the line of action of its weight is outside its base.

3 Using magnetic fields to keep things moving

3.1
1 The region around a magnet, in which a piece of iron or steel will be attracted to it
2 An electromagnet consists of a piece of insulated wire wrapped around an iron core.

3.2
1 It stays the same.
2 Zero

3.3
1 The stationary bar magnet is not cutting any magnetic field lines.
2 The direction of the induced pd is reversed.

3.4
1 Iron can be magnetised.
2 The transformer will not work, as it requires an ac supply. A cell supplies dc.

3.5
1 So that the electrical energy can be transmitted at a high pd, reducing energy wasted in the cables
2 8 V

3.6
1 X-rays
2 X-rays are absorbed by bone.

Answers to end of chapter questions
1 To stop current short-circuiting between the loops of wire
2 A line along which a plotting compass will point
3 The direction of the force
4 By increasing the strength of the magnetic field or the size of the current

5 A pd is induced across the ends of the coil.
6 By increasing the speed of movement, increasing the strength of the magnetic field, or increasing the number of turns on the coil
7 A transformer consists of two coils of insulated wire wound on an iron core.
8 A switch mode transformer operates at a much higher frequency.
9 In a sub-station before transmission to consumers
10 5
11 2 A
12 Ferrite

Answers to Practice questions

1 a The point at which the mass of the seesaw may be thought to be concentrated   (1 mark)
   b i There is a moment about the pivot from one child. For the seesaw to balance there must be an equal and opposite moment from the other child. The moment will only be opposite if the child sits on the other side of the pivot.   (2 marks)
     ii $F_1 \times d_1 = F_2 \times d_2$

$$d_2 = \frac{F_1 \times d_1}{F_2}$$

$$d_2 = \frac{460 \, \text{N} \times 2.0 \, \text{m}}{575 \, \text{N}}$$

$$d_2 = 1.6 \, \text{m} \quad (4 \, marks)$$

   c The second child has a bigger weight than the first. If they are both the same distance from the pivot the moment of the second child is bigger. So his end of the seesaw will go down.   (2 marks)

2 Marks awarded for this answer will be determined by the Quality of Written Communication (QWC) as well as the standard of the scientific response.

There is a clear, balanced and detailed explanation of how the loudspeaker makes use of the motor effect to produce a sound wave. The answer shows almost faultless spelling, punctuation and grammar. It is coherent and in an organised, logical sequence. It contains a range of appropriate or relevant specialist terms used accurately.   (5–6 marks)
There is a description of how the loudspeaker makes use of the motor effect to produce a sound wave. There are some errors in spelling, punctuation and grammar. The answer has some structure and organisation. The use of specialist terms has been attempted, but not always accurately.   (3–4 marks)
There is a brief description of how the loudspeaker makes use of the motor effect to produce a sound wave, which has little clarity and detail. The spelling, punctuation and grammar are very weak. The answer is poorly organised with almost no specialist terms and/ or their use demonstrating a general lack of understanding of their meaning.   (1–2 marks)
No relevant content.   (0 marks)

Examples of physics points made in the response:
• a coil that carries a current in a magnetic field experiences a force
• this is called the motor effect
• the current from the amplifier varies so the current in the coil varies and the force varies
• the force makes the coil move
• the coil is attached to the diaphragm so the diaphragm moves
• the diaphragm makes the surrounding air move, producing a sound wave.

3 a The iris is a ring of muscle that controls the size of the pupil and hence the amount of light entering the eye.   (2 marks)

b The light is made to converge by both the cornea and by the eye lens to form an image on the retina. *(2 marks)*

4 a There must be an alternating input supplied to the primary coil, which causes an alternating magnetic field in the iron core. This field passes through the secondary coil inducing an alternating pd across the secondary coil. *(5 marks)*

b i

$$\frac{V_p}{V_s} = \frac{n_p}{n_s}$$

$$n_p = \frac{V_p \times n_s}{V_s}$$

$$n_p = \frac{230\,V \times 30}{12\,V}$$

$$n_p = 575 \qquad \text{(3 marks)}$$

ii $V_p \times I_p = V_s \times I_s$

$$I_s = \frac{V_p \times I_p}{V_s}$$

$$I_s = \frac{230\,V \times 0.048\,A}{12\,V}$$

$$I_s = 0.92\,A \qquad \text{(2 marks)}$$

5 a Visible light enters the end of an optical fibre at an angle greater than the critical angle. The light is able to travel down the fibre by total internal reflection. The light is able to follow a curved path. The stomach can be illuminated and an image seen. *(3 marks)*

b i The small intestine is made of soft tissue. This does not absorb X-rays so will not show up on an X-ray picture. *(2 marks)*

ii Barium absorbs X-rays. So if the small intestine contains barium it will absorb X-rays and hence show up on the picture. *(2 marks)*

iii X-rays are ionising so they are damaging to cells.
The lead screen will absorb X-rays reducing the dose to the radiographer. Unlike the radiographer the patient is not regularly exposed to X-rays and infrequent exposure carries a low risk. *(4 marks)*

Glossary

A

Abdomen The lower region of the body. In humans it contains the digestive organs, kidneys, etc.

Activation energy The minimum energy needed to start off a reaction.

Active transport The movement of substances against a concentration gradient and/or across a cell membrane, using energy.

Alkali metal Elements in Group 1 of the periodic table, e.g. lithium (Li), sodium (Na), potassium (K).

Alveoli The tiny air sacs in the lungs which increase the surface area for gaseous exchange.

Amplitude The height of a wave crest or a wave trough of a transverse wave from the rest position. Of oscillating motion, is the maximum distance moved by an oscillating object from its equilibrium position.

Aorta The main artery leaving the left ventricle carrying oxygenated blood to the body.

Atomic weight The historical term that was used before relative atomic masses were defined in the 20th century. See Relative atomic mass, Ar.

Atria The small upper chambers of the heart. The right atrium receives blood from the body and the left atrium receives blood from the lungs.

B

Biconcave disc The shape of the red blood cells – a disc which is dimpled inwards on both sides.

Biofuel Fuel produced from biological material which is renewable and sustainable.

Biogas Methane produced by the fermentation of biological materials.

Bladder The organ where urine is stored until it is released from the body.

Bond energy The energy needed to break a particular chemical bond.

Breathe The physical movement of air into and out of the lungs. In humans this is brought about by the action of the intercostal muscles on the ribs and the diaphragm.

Breathing system The organ system involved in breathing: the ribs, intercostal muscles, diaphragm as well as the lungs and the tubes which bring air into the body from the outside.

Burette A long glass tube with a tap at one end and markings to show volumes of liquid, used to add precisely known amounts of liquids to a solution in a conical flask below it.

C

Capillary The smallest blood vessel that runs between individual cells. It has a wall that is only one cell thick.

Carbon neutral A process which uses as much carbon dioxide as it produces.

Centre of mass The point where an object's mass may be thought to be concentrated.

Centripetal acceleration The acceleration of an object moving in a circle at constant speed. Centripetal acceleration always acts towards the centre of the circle.

Centripetal force The resultant force towards the centre of a circle acting on an object moving in a circular path.

Charge-coupled device (CCD) Used to record and display an image.

Converging lens A lens that makes light rays parallel to the principal axis converge to (i.e. meet at) a point; also referred to as a convex lens.

Core body temperature The internal temperature of the body.

Coronary artery An artery which carries oxygenated blood to the muscle of the heart.

Critical angle The angle of incidence of a light ray in a transparent substance which produces refraction along the boundary.

CT scanner A medical scanner that uses X-rays to produce a digital image of any cross-section through the body or a three-dimensional image of an organ.

D

Deforestation Removal of forests by felling, burning, etc.

Dehydrated Lacking in water.

Deoxygenated Lacking in oxygen.

Dialysis The process of cleansing the blood through a dialysis machine when the kidneys have failed.

Dialysis machine The machine used to remove urea and excess mineral ions from the blood when the kidneys fail.

Diaphragm A strong sheet of muscle that separates the thorax from the digestive organs, used to change the volume of the chest during ventilation of the lungs.

Dioptre The unit of lens power, D.

Distillation A process which separates the components of a mixture on the basis of their different boiling points.

Diverging lens A lens that makes light rays parallel to the axis diverge (i.e. spread out) as if from a single point; also referred to as a concave lens.

Domed A curved, dome shape.

Donor The person who gives material from their body to another person who needs healthy tissues or organs, e.g. blood, kidneys. Donors may be alive or dead.

Double circulation The separate circulation of the blood from the heart to the lungs and then back to the heart and on to the body.

E

Ecology The scientific study of the relationships between living organisms and their environment.

Effort The force applied to a device used to raise a weight or shift an object.

Electromagnetic induction process of inducing a potential difference in a wire by moving the wire so it cuts across the lines of force of a magnetic field.

Embryonic stem cell Stem cell with the potential to form a number of different specialised cell types, which is taken from an early embryo.

End point The point in a titration where the reaction is complete and titration should stop.

Endoscope Optical device used by a surgeon to see inside the body. The device contains two bundles of flexible optical fibres, one used to transmit light into the body and the other used to see inside the body.

Equilibrium The point in a reversible reaction in which the forward and backward rates of reaction are the same. Therefore, the amounts of substances present in the reacting mixture remain constant.

Equilibrium The state of an object when it is at rest.

Evaporation The change of a liquid to a vapour at a temperature below its boiling point.

Exchange surface A surface where materials are exchanged.

F

Far point The furthest point from an eye at which an object can be seen in focus by the eye. The far point of a normal eye is at infinity.

Fertile A fertile soil contains enough minerals e.g. nitrates, to supply the crop plants with the all nutrients needed for healthy growth.

Fertiliser A substance provided for plants that supplies them with essential nutrients for healthy growth.

Field line See line of force.

Focal length The distance from the centre of a lens to the point where light rays parallel to the principal axis are focused (or, in the case of a diverging lens, appear to diverge from).

Functional group An atom or group of atoms that give organic compounds their characteristic reactions.

G

Gaseous exchange The exchange of gases, e.g. the exchange of oxygen and carbon dioxide which occurs between the air in the lungs and the blood.

Global warming Warming of the Earth due to greenhouse gases in the atmosphere trapping infrared radiation from the surface.

Greenhouse effect The trapping of infrared radiation from the Sun as a result of greenhouse gases, such as carbon dioxide and methane, in the Earth's atmosphere. The greenhouse effect maintains the surface of the Earth at a temperature suitable for life.

Greenhouse gas Gases, such as carbon dioxide and methane, which absorb infrared radiated from the Earth, and result in warming up the atmosphere.

Guard cells The cells which surround stomata in the leaves of plants and control their opening and closing.

H

Haemoglobin The red pigment which carries oxygen around the body.

Hard water Water in which it is difficult to form a lather with soap. It contains calcium and/or magnesium ions which react with soap to produce scum.

Heart The muscular organ which pumps blood around the body.

Herbicide Chemical that kills plants.

Homeostasis The maintenance of constant internal body conditions.

Homologous series A group of related organic compounds that have the same functional group, e.g. the molecules of the homologous series of alcohols all contain the –OH group.

Hydraulic pressure The pressure in the liquid in a hydraulic arm.

I

Immune response The response of the immune system to cells carrying foreign antigens. It results in the production of antibodies against the foreign cells and the destruction of those cells.

Immunosuppressant drug A drug that suppresses the immune system of the recipient of a transplanted organ to prevent rejection.

Industrial waste Waste produced by industrial processes.

Insulin A hormone involved in the control of blood sugar levels.

Intercostal muscles The muscles between the ribs which raise and lower them during breathing movements.

Ion-exchange column A water softener which works by replacing calcium and magnesium ions with sodium or hydrogen ions, removing the hardness.

Isotonic Having the same concentration of solutes as another solution.

K

Kidney transplant Replacing failed kidneys with a healthy kidney from a donor.

L

Line of action The line along which a force acts.

Line of force Line in a magnetic field along which a magnetic compass points; also called a magnetic field line.

Liver A large organ in the abdomen which carries out a wide range of functions in the body.

Load The weight of an object raised by a device used to lift the object, or the force applied by a device when it is used to shift an object.

Long sight An eye that cannot focus on nearby objects.

M

Magnetic pole End of a bar magnet or a magnetic compass.

Magnification The image height ÷ the object height.

Magnifying glass A converging lens used to magnify a small object that must be placed between the lens and its focal point.

Methane A hydrocarbon gas with the chemical formula CH_4. It makes up the main flammable component of biogas.

Moment The turning effect of a force defined by the equation: Moment of a force (in newtonmetres) = force (in newtons) × perpendicular distance from the pivot to the line of action of the force (in metres).

Motor effect When a current is passed along a wire in a magnetic field and the wire is not parallel to the lines of the magnetic field, a force is exerted on the wire by the magnetic field.

Mycoprotein A food based on the fungus *Fusarium* that grows and reproduces rapidly. It means 'protein from fungus'.

N

National Grid The network of cables and transformers used to transfer electricity from power stations to consumers (i.e. homes, shops, offices, factories, etc.).

Near point The nearest point to an eye at which an object can be seen in focus by the eye. The near point of a normal eye is 25 cm from the eye.

Negative pressure A system when the external pressure is lower than the internal pressure.

Non-renewable Something which cannot be replaced once it is used up.

North pole North-pointing end of a freely suspended bar magnet or of a magnetic compass.

O

Optical fibre Thin glass fibre used to send light signals along.

Oscillating motion Motion of any object that moves to and fro along the same line.

Osmosis The net movement of water from an area of high concentration (of water) to an area of low concentration (of water) along a concentration gradient.

Oxygenated Containing oxygen.

Oxyhaemoglobin The molecule formed when haemoglobin binds to oxygen molecules.

P

Parasite Organism which lives in or on other living organisms and gets some or all of its nourishment from this host organism.

Partially permeable Allowing only certain substances to pass through.

Perfluorocarbon Chemical which can be used as artificial blood.

Periodic table An arrangement of elements in the order of their atomic numbers, forming groups and periods.

Permanent hard water Hard water whose calcium and/or magnesium ions are not removed when the water is boiled, thus remaining hard.

Pesticide Chemical that kills animals.

Phloem The living transport tissue in plants which carries sugars around the plant.

Pigment A coloured molecule.

Pipette A glass tube used to measure accurate volumes of liquids.

Pivot The point about which an object turns when acted on by a force that makes it turn.

Plasma The clear, yellow liquid part of the blood which carries dissolved substances and blood cells around the body.

Platelet Fragment of cell in the blood which is vital for the clotting mechanism to work.

Positive pressure A system where the external pressure is higher than the internal pressure.

Power of a lens The focal length of the lens in metres. The unit of lens power is the dioptre, D.

Pressure Force per unit area for a force acting on a surface at right angles to the surface. The unit of pressure is the pascal (Pa).

Principal axis A straight line that passes along the normal at the centre of each lens surface.

Principal focus The point where light rays parallel to the principal axis of a lens are focused (or, in the case of a diverging lens, appear to diverge from).

Principle of moments For an object in equilibrium, the sum of all the clockwise moments about any point = the sum of all the anticlockwise moments about that point.

Pulmonary artery The large blood vessel taking deoxygenated blood from the right ventricle of the heart to the lungs.

Pulmonary vein The large blood vessel that brings blood into the left atrium of the heart from the lungs.

R

Range of vision Distance from the near point of an eye to its far point.

Real image An image formed where light rays meet.

Recipient The person who receives a donor organ.

Red blood cell Blood cell which contains the red pigment haemoglobin. It is a biconcave disc in shape and gives the blood its red colour.

Refraction The change of direction of a light ray when it passes across a boundary between two transparent substances (including air).

Refractive index Refractive index, n, of a transparent substance is a measure of how much the substance can refract a light ray.

Rehydrate To restore water to a system.

Resultant moment The difference between the sum of the clockwise moments and the anticlockwise moments about the same point if they are not equal.

Root hair cell Cell on the root of a plant with microscopic hairs which increases the surface area for the absorption of water from the soil.

S

Scale (limescale) The insoluble substance formed when temporary hard water is boiled.

Scum The precipitate formed when soap reacts with calcium and/or magnesium ions in hard water.

Sewage A combination of bodily waste, waste water from homes and rainfall overflow from street drains.

Short sight An eye that cannot focus on distant objects but can focus on near objects.

Simple pendulum A pendulum consisting of a small spherical bob suspended by a thin string from a fixed point.

Soapless detergent A cleaning agent that does not produce scum when used with hard water.

Soft water Water containing no dissolved calcium and/or magnesium salts, so it easily forms a lather with soap.

Solute The solid which dissolves in a solvent to form a solution.

South pole South-pointing end of a freely suspended bar magnet or of a magnetic compass.

Split-ring commutator Metal contacts on the coil of a direct current motor that connects the rotating coil continuously to its electrical power supply.

Stent A metal mesh placed in the artery which is used to open up the blood vessel by the inflation of a tiny balloon.

Step-down transformer Electrical device that is used to step down the size of an alternating voltage.

Step-up transformer Electrical device that is used to step up the size of an alternating voltage.

Strong acids Acids that ionise completely in aqueous solutions.

Sustainable food production Methods of producing food which can be sustained over time without destroying the fertility of the land or ocean.

Switch mode transformer A transformer that works at much higher frequencies than a traditional transformer. It has a ferrite core instead of an iron core.

T

Temporary hard water Hard water which is softened when it is boiled.

Thermoregulatory centre The area of the brain which is sensitive to the temperature of the blood.

Time period Time taken for one complete cycle of oscillating motion.

Titration A method for measuring the volumes of two solutions that react together.

Total internal reflection Occurs when the angle of incidence of a light ray in a transparent substance is greater than the critical angle. When this occurs, the angle of reflection is equal to the angle of incidence.

Trachea The main tube lined with cartilage rings which carries air from the nose and mouth down towards the lungs.

Transformer Electrical device used to change an (alternating) voltage. See also Step-up transformer and Stepdown transformer.

Transfusion The transfer of blood from one person to another.

Transition element Element from the central block of the periodic table. It has typical metallic properties and forms a coloured compound.

Transition metal See Transition element.

Transpiration stream The movement of water through a plant from the roots to the leaves as a result of the loss of water by evaporation from the surface of the leaves.

Transport system A system for transporting substances around a multicellular living organism.

Type 1 diabetes Diabetes which is caused when the pancreas cannot make insulin. It usually occurs in children and young adults and can be treated by regular insulin injections.

U

Ultrasound wave Sound wave at frequency greater than 20 000 Hz, which is the upper limit of the human ear.

Urea The waste product formed by the breakdown of excess amino acids in the liver.

V

Vacuum An area with little or no gas pressure.

Valve Structure which prevents the backflow of liquid, e.g. the valves of the heart or the veins.

Vena cava The large vein going into the right atrium of the heart carrying deoxygenated blood from the body.

Ventilated Movement of air into and out of the lungs.

Ventricles The large chambers at the bottom of the heart. The right ventricle pumps blood to the lungs, the left ventricle pumps blood around the body.

Villus A finger-like projection from the lining of the small intestine which increases the surface area for the absorption of digested food into the blood.

Virtual image An image, seen in a lens or a mirror, from which light rays appear to come after being refracted by the lens or reflected by the mirror.

W

Weak acids Acids that do not ionise completely in aqueous solutions.

Wilting The process by which plants droop when they are short of water or too hot. This reduces further water loss and prevents cell damage.

X

X-ray High energy wave from the part of the electromagnetic spectrum between gamma rays and ultraviolet waves.

Xylem The non-living transport tissue in plants, which transports water around the plant.

Acknowledgements

The authors and the publisher would like to thank the following for permission to reproduce material:

Text

pp30, 31: *AQA Examination questions are reproduced by permission of AQA Education (AQA).*

Images

p1: iStockphoto

Biology (B3)

p5: BSIP Laurent/Trunyo/Science Photo Library; p11: National Cancer Institute/Science Photo Library; p12: Martin Leigh/Oxford Scientific Film/Photolibrary/GettyImages; p23: Fotolia; p25: Victor de Schwanberg/Science Photo Library; p26: NREL/US Department of Energy/Science Photo Library; p27, intensively reared chickens: National Geographic/Getty Images, free-range chickens: The Image Bank/Getty Images; p30: Wim Van Egmond, Visuals Unlimited/Science Photo Library.

Chemistry (C3)

p32, Dmitri Mendeleev: CCI Archives/Science Photo Library, Mendeleev's periodic table: Science Photo Library; p35: Tony Craddock/Science Photo Library; p36: Andrew Lambert Photography/Science Photo Library; p38, scum in sink: Martyn F. Chillmaid/Science Photo Library, lime scale in heating systems: iStockphoto; p39: Pink Sun Media/Alamy; p40, iStockphoto; p46, hydrogen re-fuelling station: Dane Andrew/CorbisNews, London bus: Toby Melville/Reuters/Corbis; p48: David Taylor/Science Photo Library; p49, barium sulfate: Charles D. Winters/Science Photo Library, silver chloride, silver bromide and silver iodide: Science Photo Library; p50: Andrew Lambert Photography/Science Photo Library; p52, genetic analysis: Patrick Dumas/Eurelios/Science Photo Library, forensic drug analysis: Patrick Landmann/Science Photo Library; p58: iStockphoto; p59, Lawrence Migdale/Science Photo Library; p60, iStockphoto.

Physics (P3)

p79: Cordelia Molloy/Science Photo Library.

Every effort has been made to trace the copyright holders but if any have been inadvertently overlooked the publisher will be pleased to make the necessary arrangements at the first opportunity.